Alternative Energy: Powering the Future

Alternative Energy: Powering the Future

Edited by David McCartney

SYRAWOOD
PUBLISHING HOUSE

New York

Published by Syrawood Publishing House,
750 Third Avenue, 9th Floor,
New York, NY 10017, USA
www.syrawoodpublishinghouse.com

Alternative Energy: Powering the Future
Edited by David McCartney

International Standard Book Number: 978-1-68286-608-5 (Hardback)

Cataloging-in-Publication Data

Alternative energy : powering the future / edited by David McCartney.
    p. cm.
Includes bibliographical references and index.
ISBN 978-1-68286-608-5
1. Renewable energy sources. 2. Power resources. I. McCartney, David.
TJ808 .A48 2018
333.794--dc23

TABLE OF CONTENTS

Preface..VII

Chapter 1   **Theoretical potential and utilization of renewable energy in
            Afghanistan**..1
            Gul Ahmad Ludin, Mohammad Amin Amin, Assadullah Aminzay,
            and Tomonobu Senjyu

Chapter 2   **Evaluating the potential of renewable diesel production from
            algae cultured on wastewater: techno-economic analysis and life
            cycle assessment**..20
            Ankita Juneja and Ganti S. Murthy

Chapter 3   **Design of an off-grid hybrid PV/wind power system for remote
            mobile base station**..39
            Mulualem T. Yeshalem and Baseem Khan

Chapter 4   **Analyses of optimum generation scenarios for sustainable power
            generation in Ghana**...56
            Albert K. Awopone and Ahmed F. Zobaa

Chapter 5   **Energy security and competition over energy resources in Iran and
            Caucasus region**...72
            Shahrouz Abolhosseini, Almas Heshmati
            and Masoomeh Rashidghalam

Chapter 6   **Performance and emission reduction potential of micro-gasifier
            improved through better design**...87
            Kamil Dino Adem and Demiss Alemu Ambie

Chapter 7   **Assessment of offshore wind power potential in the Aegean and
            Ionian Seas based on high-resolution hindcast model results**.............101
            Takvor Soukissian, Anastasios Papadopoulos, Panagiotis Skrimizeas,
            Flora Karathanasi, Panagiotis Axaopoulos, Evripides Avgoustoglou,
            Hara Kyriakidou, Christos Tsalis, Antigoni Voudouri,
            Flora Gofa and Petros Katsafados

Chapter 8   **Influence of design parameters on the structural and fatigue
            behaviors of a floating point wave energy converter**...................123
            Pedro J. B. F. N. Beirão, Cândida M. S. P. Malça and
            Raimundo P. Felismina

Chapter 9   **Global changes in total and wind electricity (1990–2014)**.............138
            María del P. Pablo-Romero and Rafael Pozo-Barajas

Chapter 10   **Optimization of regular offshore wind-power plants using a non-discrete evolutionary algorithm**..161
Angel G. Gonzalez-Rodriguez, Manuel Burgos Payan,
Jesús Riquelme Santos and Javier Serrano Gonzalez

Chapter 11   **Effect of the storage condition of microalgae on hydrochar lipids and direct esterification-transesterification of hydrochar lipids for biodiesel production**...181
Vo Thanh Phuoc and Kunio Yoshikawa

Chapter 12   **Oil extraction from plant seeds for biodiesel production**...196
Yadessa Gonfa Keneni and Jorge Mario Marchetti

Permissions

List of Contributors

Index

PREFACE

The rise in pollution is posing a threat to our environment. Technologies are being developed across the globe which are environment-friendly. Renewable resources are resources that can be renewed and have low carbon imprint. Some of the natural resources used for renewable energy include sunlight, tides, wind and rain. The aim of this book is to present topics that have transformed this discipline and aided its advancement. From theories to research to practical applications, case studies related to all contemporary topics of relevance to this field have been included herein. It will serve as a reference to a broad spectrum of readers including environmentalists, engineers, researchers, professionals and students associated with this area of study.

After months of intensive research and writing, this book is the end result of all who devoted their time and efforts in the initiation and progress of this book. It will surely be a source of reference in enhancing the required knowledge of the new developments in the area. During the course of developing this book, certain measures such as accuracy, authenticity and research focused analytical studies were given preference in order to produce a comprehensive book in the area of study.

This book would not have been possible without the efforts of the authors and the publisher. I extend my sincere thanks to them. Secondly, I express my gratitude to my family and well-wishers. And most importantly, I thank my students for constantly expressing their willingness and curiosity in enhancing their knowledge in the field, which encourages me to take up further research projects for the advancement of the area.

Editor

Theoretical potential and utilization of renewable energy in Afghanistan

Gul Ahmad Ludin [1,*], **Mohammad Amin Amin** [2], **Assadullah Aminzay** [2], **and Tomonobu Senjyu** [1]

[1] Electrical and Electronics Engineering Department, University of the Ryukyus, Okinawa, Japan
[2] Electrical Power Engineering Department, Kabul Polytechnic University, Kabul, Afghanistan

* **Correspondence:** Email: gulahmadludin@gmail.com

Abstract: Nowadays, renewable energy is gaining more attention than other resources for electricity generation in the world. For Afghanistan that has limited domestic production of electric power and is more dependent on the unstable imported power from neighboring countries which pave the way to raise the cost of energy and increased different technical and economic problems. The employment of renewable energy would not only contribute to the independence of energy supply but also can achieve the socio-economic benefits for the country which is trying to rebuild its energy sector with a focus on sustainable energy for its population. From a theoretical point of view, there is a considerable potential of renewable energies such as solar energy, wind power, hydropower, biomass and geothermal energy available in the country. However, despite the presence of widespread non-agricultural and non-residential lands, these resources have not been deployed efficiently. This paper assesses the theoretical potential of the aforementioned types of renewable energies in the country. The study indicates that deployment of renewable energies can not only supplement the power demand but also will create other opportunities and will enable a sustainable energy base in Afghanistan.

Keywords: solar; wind; biomass; hydro; geothermal; Afghanistan

1. Introduction

Afghanistan is a mountainous and landlocked country which is located in South-Central Asia, north and west of Pakistan, east of Iran. It is located between latitude $29°35'$ and $38°40'$ degrees north and longitude $60°31'$ and $75°00'$ east. The total land area is 652,864 km^2 with arid and semiarid climate and cold winters and hot summers [1]. The total population is 31,627,506 in 2014 with more than 75% (27.5 million people) live in rural areas [2,3].

The country is facing with significant environmental problems; such as limited natural fresh water resources, inadequate supplies of potable water, soil degradation, overgrazing, deforestation (much of the remaining forests are being cut down for fuel and building materials), desertification and air and water pollution.

Access to electricity is a higher development priority and also the second priority after the rule of law, for Afghanistan. Most of the infrastructures of the country, specifically its national grid (existed power plants, substations, transmission and distribution networks) were destroyed and demolished during the last 30 years of civil war and instability.

Besides, as energy is the backbone of the national economy of each country and also, living standard of each country's citizens is indicated by GDP of the country which is claimed to be about 90% dependent on per capita energy consumption. Afghanistan with its $687 GDP per capita and 78.21 kWh electricity consumption per capita (195 kWh per annum) in 2014, ranks among the lowest in the world [4,5,6].

The main reasons that only 9% rural and 30% of the whole country's population has access to electricity are: firstly, the majority of the population live in the rural area, and expansion of the existing transmission network will be very expensive. Secondly, even if the first is to be accomplished the major portion (78%) of supplied electricity in Afghanistan is the power imports from neighboring countries which is claimed to be very expensive and unaffordable for its citizens especially for those who live in the suburbs with the lowest income mainly from agriculture and livestock. The main concern is the resource, energy generation from fossil fuel based plants, due to drawbacks that they have, seems not relevant to the 21st century anymore. As NASAs Earth Observatory explained that global warming was the unusually rapid increasing in Earths average surface temperature over the past century primarily due to the greenhouse gasses (GHGs) released as people burn fossil fuels. Besides, according to the world health organization (WHO), annually 7 million deaths are due to the air pollution in the world [7]. It has been estimated that without taking any plan to reduce GHGs emissions, the global average surface temperature will rise by 1.8–4 °C by the end of this century and will increase the earth's temperature above the threshold value which could cause irreversible and possibly catastrophic changes. Finally, such Projected global warming in this century is likely to trigger serious consequences for humanity and other life forms [8].

In Afghanistan, large coal-burning power plants will considerably increase the greenhouse gas emission. According to the world bank database, Afghanistan's GHGs emissions are less than 0.1 tons/capita ranks among the lowest in the world [9].

Based on the Afghan Rural Renewable Energy Strategy, 85% of primary energy demand in Afghanistan is met by traditional biomass mainly wood and dung, which is used for cooking and heating purposes [10,11]. The total internal power plant installed generation as depicted in Figure 1, in the country is mainly hydro power stations, diesel generators and thermal power plants [12].

Since 2001, the main focus of endeavors has been on reconstruction and expansion of the national power grid. However, given that the national grid is being developed almost from scratch, it is accepted that there are swathes of the country as depicted in Figure 2, that the national grid will not be able to serve them in the next 20 years [11,13]. It is also possible that supply constraints will mean that some grid-connected areas are unlikely to receive a sufficient and reliable power supply.

On the other hand, according to the Afghanistan power sector master plan, the total power demand is forecasted to reach 3500 MW excluding losses in 2032.

In order to come out from the challenges mentioned above and supplement the energy demand of the country, the deployment and development of internal resources are required.

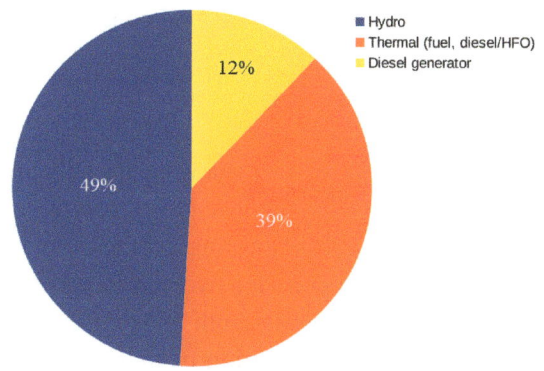

Figure 1. Afghanistan power plants installed capacity [12].

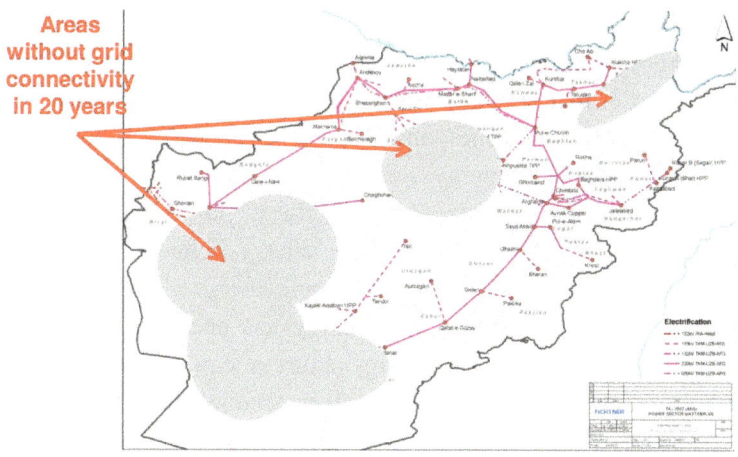

Figure 2. The areas without grid connectivity in Afghanistan [11] & [13].

Afghanistan has significant renewable energy resources especially solar, wind, hydro, biomass and geothermal in the country. Utilization of these resources could be effectively way not only to meet the increasing demand but also will create more opportunities such as poverty reduction and creating different kind of jobs. Besides, if compared to the conventional energy sources, the environmental impacts from renewables are considerably small [14]. In addition, it may be more viable for some areas which have significant renewable energy potential, to serve by local renewables than to provide grid-connected supply at subsidized tariffs that do not cover the costs of service. Moreover, the ministry of energy and water of Afghanistan also realized the importance of renewable energy deployment particularly for rural electrification in the country. Thus, the ministry has established an independent organization in 2009, called renewable energy department [15].

This paper presents the theoretical resource potential and utilization of renewable energy in the country. All the data is gathered from energy sector's related organizations in the country, energy sector master plan, papers, reports, articles, magazines and websites. The paper is organized as follows: Section 2 is dedicated for literature review, Section 3 briefly describes the Afghan energy system, Section 4 is organized for assessment of renewable energy potential in the country. Section 5 is dedicated for policy recommendations. Finally, results and conclusion are presented in section 6.

2.   Literature Review

In order to clarify the role of renewable energy as an eco-friendly and user-friendly resource for supplying the energy demand and other socio-economic development purposes, literature review on the renewable energy development of several developing countries including Afghanistan has been done in this section. In [16] a study has been done for two communities in two phase on the benefits of renewable energy technologies and their impacts both economically and socially in Afghanistan. The authors found and indicated that, by electrification of these two communities with renewables (PV and micro-hydro technologies) the socio-economic achievements such as job creation, personal security, family interaction, learning conditions for children, entertainment opportunities, access to information, improving health conditions, household income and finally energy sustainability and affordability have increased. In another study [18], author investigated the resource potential of solar and wind power in Balkh and Herat, two most promising provinces of Afghanistan. It was founded that Solar PV and wind power plants in these two provinces have the potential to reduce the load on all types of domestic generation schemes and imports, especially if both resources could have utilized. Besides, in the Afghanistan energy sector master plan [11], a German company Fichtner recommended that distributed hybrid wind, solar and diesel power plants and off-grid solar home systems are the most promising options to meet the energy demand mainly in rural areas of the country.

In [17] the resource potential of renewable energy has been assessed in Pakistan. The study results clearly demonstrate that renewable energy sources can supplement the energy needs of Pakistan and can provide a sustainable energy base. It was concluded that exploitation of renewable resources can help Pakistan in developing a sustainable energy base and at the same time enhance energy security of the country by reducing its dependence on imported fossil fuels. Iran is also one of the developing countries which is the biggest producer of oil and gas in the world that are used in diverse industrial section such electric power plants. The CO_2 emission is considerable and placed the country among the top ten emitting countries. Air pollution in Iran's large cities is the most critical energy-related pollutant. The government is trying to follow the energy sustainability in the Iran by development of renewable energy technology. The target is set to 20000 MW (10% of the required electricity) by 2025. Due to the various renewable energy policy challenges, the Iran's energy system looks far from sustainability [19]. Similarly, Turkmenistan possesses a huge potential for wind and solar electric energy, but rich gas and oil deposits have not encouraged development of this sector. Electricity generation in Turkmenistan is mainly from thermal power plants. In policy documents on climate change mitigation, the Turkmen Government aims to increase renewable energy generation, but a legislative framework to promote and support investment in renewable energy does not yet exist. On the other hand, the National Program of Turkmenistan identifies renewable energy technologies as contributing to meeting several targets, including environmental protection and sustainability [20]. Tajikistan has one of the world's largest hydropower potential which provides more than 90% of the country's energy demand. The main policy addressing climate change is the national action plan for climate change mitigation (2003). Within this policy, the development of renewable energies, including small scale hydropower, are in top priorities.

Besides, in the target program for the widespread use of renewable energy sources, the use of renewable energy is recognized as a mean of achieving main development goals. The second national communication to the United Nations Framework Convention on Climate Change (UNFCCC) stated that Tajikistan has a large potential for development of small hydropower, and estimates that the use of existing technical potential for small hydropower would lead to an annual reduction of 5m to 6m tonnes of CO_2 emissions. There are a number of laws and policies that support the development of renewables [21].

In 2007, the government adopted the special program for renewable energy sources use in Tajikistan for 2007–2015. This program introduced a set of measures to create a production base and infrastructure for wider use of renewable sources of energy: solar, wind, biomass, small hydro and geothermal. The purpose of the program was to develop and deploy technologies for electricity and heating generation from renewable energy sources; to raise living standards; to reduce the use of non-commercial biomass fossil fuels; to train qualified personnel; to develop remote, off-grid areas; and to contribute to environmental protection [22].

A research has been done about renewable energy potential in Bangladesh in [23]. The authors concluded that, for supplying the increasing demand and also access to modern, reliable, affordable, and sustainable energy services, utilization of renewable energies would play a positive role in economic and social progress in Bangladesh.

In [24], the energy policy of several Asian and African developing countries are compared by presenting their corresponding GDP and their preset targets for the integration of RE technologies. It is indicated that the circumstances of each country and their standard of living is directly reflected in the modesty or sophistication of their future energy plans and a suitable and possible development option is using Renewable Energy Technologies (RETs). The main issue preventing the development and implementation of such projects is the lack of clear policies and legal frameworks set by the governments in addition to extremely complicated political issues and conflicts, and usually limited financial means. It is also stated that major help should be offered by developed countries, in order for African nations to overcome the energy poverty issue by development of renewable energies.

3. Afghan Energy System

Currently, Afghanistan does not possess any grid-scale renewable energy assets. While gifted with solar, wind and biomass resource potential, the country relies primarily on power imports, followed by hydro and thermal power plants and diesel generation. DABS (Da Afghanistan Breshna Sherkat) is the state-owned provider of electricity in Afghanistan and has the possession of generation, transmission and distribution services, operation and maintenance of assets, sales of electricity and revenue collection inside the country. Figure 3 depicts the general state of electricity market that matches the purchasing agency one of the four famous market competition models of Hunt and Shuttle worth [25], which shows the possible first step toward the introduction of competition in the electricity supply industry. The integrated utility (DABS) no longer has the possession of all the generation companies. Instead independent power producers (IPPs) as power imports, are connected to the utility that acts as a purchasing agency.

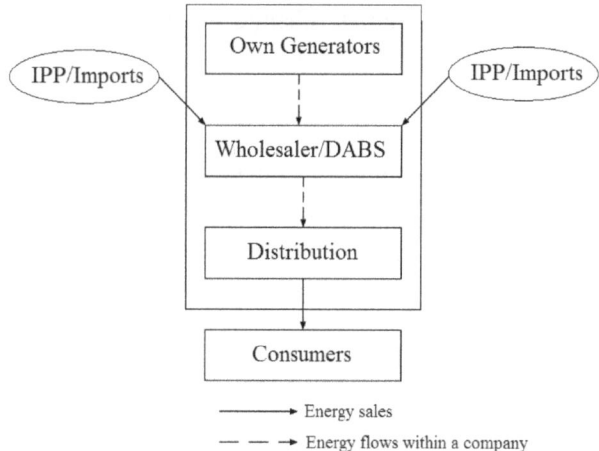

Figure 3. General state of the power market [25].

This model has both the advantages and disadvantages. This case of trade introduces some competition between IPPs without the expense of setting up a competitive market. Unlikely, this model does not discover a cost reflective price in the same way that a free market does. Moreover, it shows that the market is most dependent on IPP (Importers for the case of Afghanistan).

However, in the past few years, due to the population growth, economic development, rapid urbanization, and industrial development of the country, electricity consumption has grown rapidly in Afghanistan. The enhanced availability of imported power from neighboring countries has been the major driver in this growth. This is the result of the construction of North East Power System (NEPS), transmission system, that interconnects Uzbekistan and Tajikistan to Afghanistan and the expansion of the distribution networks. Besides, 105 MW thermal power plant at Tarakhil was commissioned in 2009. Furthermore, rehabilitation of hydro stations in the country mainly serving Kabul i.e. Naghlu, Mahiper and Sarobi is the mentionable factor in this growth. Connection rates increased from 7% in 2003 to 30% in 2014, and it is projected to reach 83% by the year 2032 [3,18,26].

DABS's statistics indicated that the total number of consumers increased from about 200,000 in 2003 to 800,000 in 2010 [27]. Figure 4 shows the net demand projection of electricity for Afghanistan from 2011 to 2032 [11]. As it is evident, in the high scenario the total net demand is expected to increase by a factor of 8.8 from 2011 and would reach to 19,474 GWh by 2032 or 22% higher than in the base case. Figure 5 shows the projection of peak load development from 2011 to 2032 for the entire Afghanistan which increases by factor of 5.7 and would reach to 3502 MW in 2032 [11]. Supplying the demand is bounded not only by limited domestic generation but also it is repressed by national instability, unstable imports and unreliable national electrical networks.

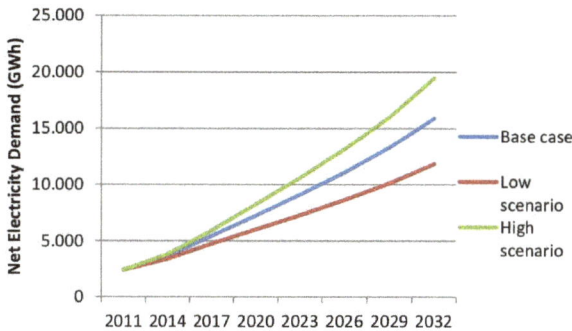

Figure 4. Development of net electricity demand in Afghanistan [11].

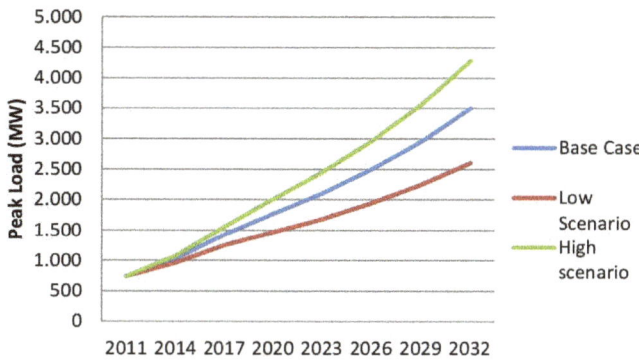

Figure 5. Development of peak load in Afghanistan [11].

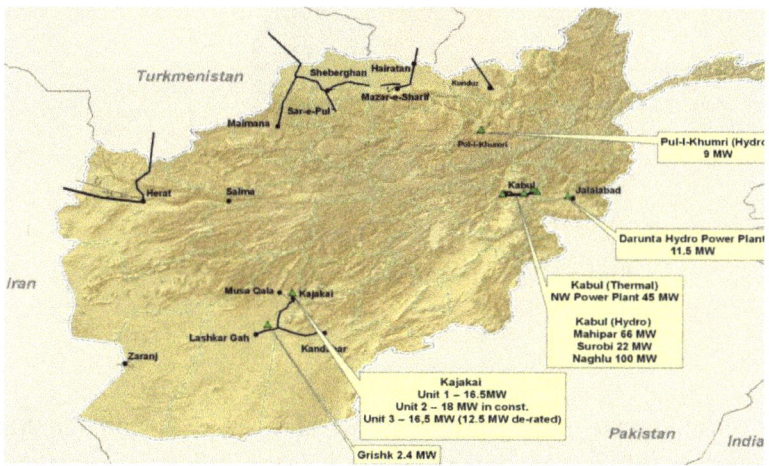

Figure 6. Locations of hydro power plants in Afghanistan [10].

Of the estimated 5 million MWh of electricity consumed in Afghanistan in 2013, 78% was imported from neighboring countries; 345 MW from Uzbekistan, 300 MW from Turkmenistan, 300 MW from Tajikistan and 70 MW from Iran [9]. Domestic generation is dominated by hydropower plants which totally make 258 MW of the total supplied power. Figure 6 shows the locations of more than 1MW existing hydro power stations throughout the country [10]. The remaining electricity demand is met by internal local generation from micro hydropower, solar, wind and small diesel Generators. The total installed capacity of Afghanistan is projected 3600 MW by 2032, where the Share of Afghan own generation will be 67% of the total capacity. As 83% of total population will have access to electricity in 2032, the need for power imports is decreased to only 16% of the entire demand [28]. On the other hand, given this supply scenario, Afghanistan is currently embarking on ways to attract clean and sustainable energy developers and interested in renewable Independent Power Producers (IPPs).

4.   Renewable Energy Potential

Theoretical or total potential is the highest level of resource potential that only considers the restrictions with respect to natural and climatic parameters [29]. The evidence shows that from a theoretical point of view, there is a considerable potential of renewable energy resources especially solar, wind, hydro, biomass and geothermal available in the country. Although, still there is no grid-scale renewable energy excluding hydropower in the country, considerable works have been done in this field. Figure 7 shows the amounts and capacities of renewable energy projects either ongoing or accomplished in the country by 2015 [30]. This section briefly assesses the available theoretical potential of types mentioned above of renewable energy sources in the country.

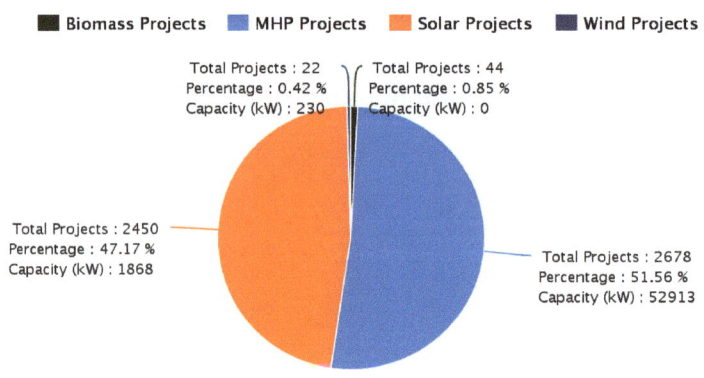

Figure 7. Renewable energy projects [30].

4.1. Solar energy

Solar energy constitutes the major portion of renewable energy potential in the country. Besides, as it is clean, abundant, offers zero input cost and distributed throughout the country, it is a prerequisite sustainable source especially for remote off-grid locations. Afghanistan with having about 300 days of the sunshine per year and estimated 6.5 kWh/m^2/day of theoretical solar energy density, has significant solar energy potential which can be tapped using solar thermal and PV technologies [5,31]. From a technical point of view, the plants can be both off-grid and micro grids for residential and industrial customers own use or potentially large solar plants connected to the grid. However, there has been no evidence of any plans to deploy large solar power plants, but the only plan in hand is of national solidarity program (NSP) and rural electrification. More than 2000 individual solar projects have been implemented through this program by 2014. Most of these projects are for producing electricity for lighting in some schools, hospitals, police buildings, mosques and some residential areas in the villages. The technology has been used, is installation more than 100,000 PV systems mostly of less than 50 Wp capacity [31].

Afghanistan solar radiation 10 km resolution annual map for PV technology, prepared by national renewable energy laboratory (NREL) is shown in Figure 8. As it is evident that, the major spots of solar resources are located in southern, south-eastern and south-western areas of the country. According to the statistics of ministry of energy and water, the total feasible potential of solar energy is about 222 GW throughout the country [31]. Table 1 shows the solar radiation and the total and feasible potential of solar power per each province in the country [30]. If we assume that, only 2% of the total land (652,864 km^2) is used considering 300 sun shining days, 12 hours per day and average 6.5 kWh/m^2/day energy density, then the maximum feasible potential can be given as Eq. 1. Now, if poly-crystalline silicon PV cells with having 16% conversion efficiency are used, then it will produce 3,768.32 MW power. This is still a huge capacity and even it can satisfy the future power demand of the country.

$$P_{\max} = \frac{6.5kWh \times 13057.6 \times 10^6\, m^2}{3600h \times m^2} = 23.575GW \tag{1}$$

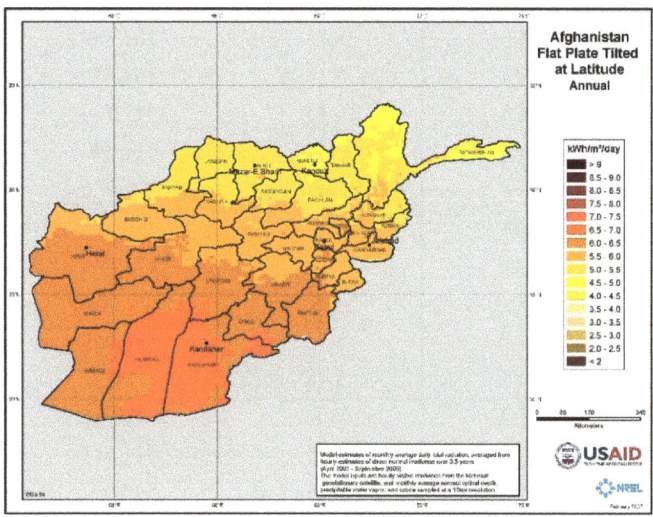

Figure 8. Annual solar radiation map of Afghanistan [11,32].

4.2. Wind energy

Nowadays, wind resource is also gaining attention for electricity generation, because it is free, relatively low capital cost and pollution free source. It is estimated that wind power in many countries is already competitive with fossil fuel and nuclear power if social/environmental costs are considered [33]. Furthermore, according to the European wind energy association (EWEA), wind energy which needs strong global commitment, will make about 12% of the world total energy supply by 2020 [34].

Table 1. Solar resource potential per each province [30,31].

No.	Provinces	Provinces area (km^2)	Solar radiation (kWh/m^2/day)	Total potential (MW)	Feasible potential (MW)
1	Badakhshan	44,836	5	3,736,325	3,736
2	Badghis	20,794	6.15	2,131,385	5328
3	Baghlan	18,255	5.05	1,536,479	1,536
4	Balkh	16,186	4.3	1,160,018	2,900
5	Bamyan	18,029	6.2	1,863,017	1,863
6	Daykundi	17,501	6.55	1,910,570	1,911
7	Farah	49,339	6.6	5,427,301	27,137
8	Faryab	20,798	5.4	1,871,784	4,679
9	Ghazni	22,460	6.2	2,320,867	5,802
10	Ghor	36,657	6.9	4,215,601	10,539
11	Helmand	58,305	6.85	6,656,499	33,282
12	Heart	55,869	6.13	5,707,898	28,539
13	Jawzjan	11,292	4.74	892,029	2,230
14	Kabul	4,524	5.73	432,032	432
15	Kandahar	54,845	6.8	6,215,710	31,079
16	Kapisa	1,908	5.75	182,850	183
17	Khost	4,235	5.15	363,530	364
18	Kunar	4,926	5.45	447,436	447
19	Kundoz	8,081	3.8	511,790	1,279
20	Laghman	3,978	5.08	336,796	842
21	Logar	4,568	5.93	451,471	451
22	Nangarhar	7,641	5.3	674,964	1,687
23	Nimroz	42,410	6.4	4,523,680	22,618
24	Nooristan	9,267	5.75	888,059	888
25	Paktia	5,583	5.48	509,932	510
26	Paktika	19,516	6.2	2,016,643	5,042
27	Panjshir	3,772	5.95	374,017	374
28	Parwan	5,715	5.75	547,697	548
29	Samangan	13,438	5.2	1,164,609	2,912
30	Sar-i-pul	16,386	6.05	1,652,215	4,131
31	Takhar	12,458	4.9	1,017,387	2,543
32	Urozgan	11,474	6.83	1,306,090	6,530
33	Wardak	10,348	6.05	1,043,454	1,043
34	Zabul	17,472	6.5	1,892,778	9,464
Total		**652,864**	**6.5**	**65,982,912**	**222,652**

Theoretically, wind energy sources in Afghanistan have been assessed by NREL as depicted in Figure 9. As it is evident that, the wind resource major areas in the country include western Afghanistan especially, Nimroz and Herat, northwestern Farah, northeastern areas especially Balkh,

Takhar, wind corridor areas including near Jabalsaraj, Sarobi and Tirgari, eastern Afghanistan particularly Qalat, Gadamsar, Walakhor, Golestan, and Gorzanak and elevated mountain summits in central/southern and northern Afghanistan. Also, as depicted in the figure, there is a tremendous potential of class 4 with 400–600 W/m^2 power density and wind speed of 6.8–7.7 m/s which is suitable for utility-scale applications. The total theoretical potential is estimated about 158000 MW (with 5 MW/km^2), and the total feasible potential is 66726 MW [15]. This means, if installed capacity is assumed 5 MW/km^2 and also if 5% of the total land (31,611 km^2) is used as stated in [25], then theoretical or total potential, of wind energy in the country can be calculated as:

$$P_{max} = 5\frac{MW}{km^2} \times 31611 km^2 = 158055 MW \tag{2}$$

However, almost 12% of Afghanistan land area has been claimed to have Class 3 or better wind resource. Table 2 shows the amount of total and feasible potential per each province of the country. Currently, there is not any utility-scale wind generation available in the country, though totally six projects with 230 kW installed capacity are completed by 2015. The first power generating wind farm with ten turbines and total installed capacity of 100 kW was built in Panjshir province by 2008 [30].

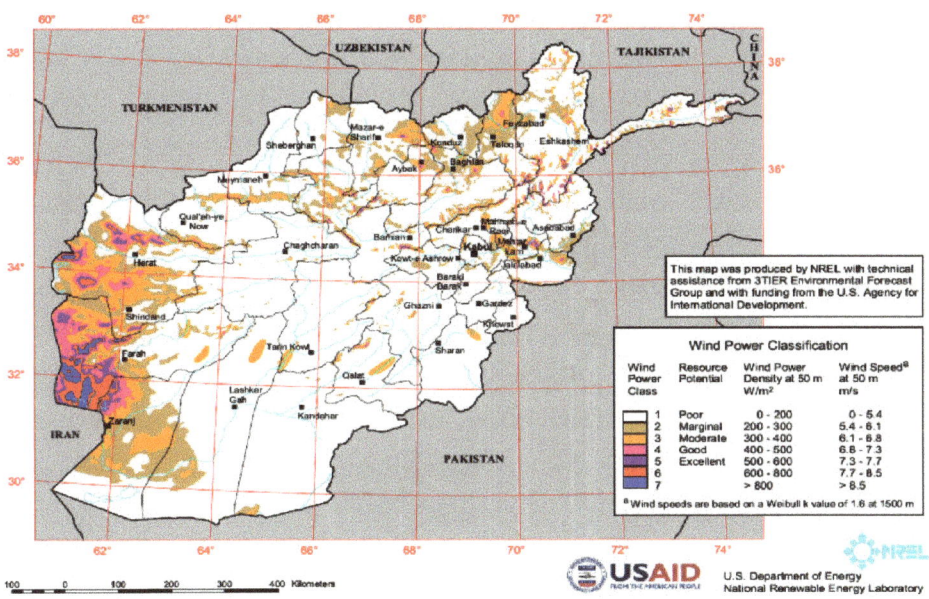

Figure 9. Wind speed map of Afghanistan [11].

Table 2. Wind power potential per each province [30,31].

No.	Provinces	Provinces area (km²)	Windy area (km²)	Total potential (MW)	Feasible potential (MW)
1	Badakhshan	44,836	1,428	3,314	331
2	Badghis	20,794	410	762	191
3	Baghlan	18,255	1,064	2083	208
4	Balkh	16,186	1,689	3,145	786
5	Bamyan	18,029	100	240	24
6	Daykundi	17,501	-	-	-
7	Farah	49,339	19,270	61,353	30,677
8	Faryab	20,798	560	1,008	252
9	Ghazni	22,460	93	191	48
10	Ghor	36,657	160	336	84
11	Helmand	58,305	1,040	1872	936
12	Heart	55,869	14,694	36,947	18,473
13	Jawzjan	11,292	95	171	43
14	Kabul	4,524	230	414	41
15	Kandahar	54,845	130	234	117
16	Kapisa	1,908	450	810	81
17	Khost	4,235	-	-	-
18	Kunar	4,926	40	72	7
19	Kundoz	8,081	180	324	81
20	Laghman	3,978	460	1020	255
21	Logar	4,568	-	-	-
22	Nangarhar	7,641	300	582	146
23	Nimroz	42,410	10,130	21,450	10,725
24	Nooristan	9,267	90	-	-
25	Paktia	5,583	-	-	-
26	Paktika	19,516	220	396	99
27	Panjshir	3,772	80	180	18
28	Parwan	5,715	705	1269	127
29	Samangan	13,438	503	1,064	266
30	Sar-i-pul	16,386	385	729	182
31	Takhar	12,458	2,547	4,795	1,199
32	Urozgan	11,474	550	990	495
33	Wardak	10,348	80	180	18
34	Zabul	17,472	860	1,632	816
Total		**652,864**	**58,543**	**158,055**	**66,726**

4.3. Biomass potential for electricity generation

Solid biomass comprises the largest portion of Afghanistan primary energy supply which could be summarized as agricultural resources such as crop residues, animal wastes, forest resources and urban wastes. According to the ministry of energy and water database, currently 44 biomass projects are completed, but they are used for non-electricity purposes in the country. From the above available listed resources, crop residues which represent a significant component of the solid biomass are using as animal feed, farm applications, construction material and hugely used for cooking purposes. Therefore, currently, they are not available for electricity production.

According to Milbrandt and Overend (2011), animal manure is also a major feature of the Afghan agriculture. About 79% of rural households and 94% of Kuchi population own some kinds of livestock. However, at the moment it is used only for cooking and heating purposes. On the other hand, forest resources are limited in the country, but still they represented a significant component of the solid biomass fuels and used for cooking and space heating [35]. Besides, municipal solid waste (MSW) in the urban areas of the country can be considered as an important source of electricity and heat production using waste-to-energy (WTE) combustion technologies.

Figure 10 demonstrates the potential of power production from biomass resources in Afghanistan in 2015. Table 3 indicates the amount of crop residues, animal manures and municipal solid waste and expected electricity generation potential per each province of the country in 2014. For conversion to electricity, the coefficients used are 550 kWh/tone of organic waste, 155 kWh/tone of animal manure and 4170 kWh/tone of crop residues, respectively [35].

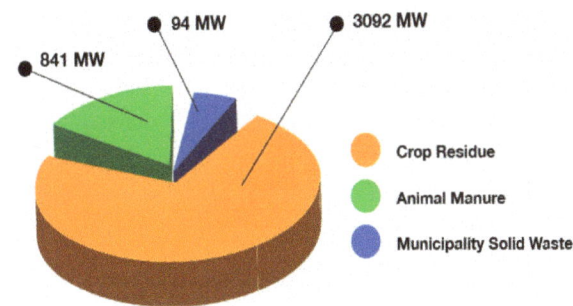

Figure 10. Potential of biomass resources by MW [30].

4.4. Hydro power

The major portion of electricity production is comprised of large hydro power plants inside the country. Theoretical potential of hydro energy is estimated 23, 310 MW (258 MW operating, 56 MW under construction and 22998 MW still untapped), where totally more than 5,000 MW is claimed to be technically possible to tap [11,36]. Table 4 indicates the summary of hydropower potential by each zone in Afghanistan. On the other hand, micro hydro power which is the most promising source especially for the rural electrification, has been developed significantly. According to the latest database of MEW and MRRD, the total installed capacity of micro hydropower plants (MHP) in Afghanistan is 36.907 MW with further 5.845 MW are under construction and 10.144 MW is surveyed by the various organization including MEW and MRRD associated programs. Totally, 125 sites have been identified for micro-hydro resource development with the potential to generate more than 600 MW of electric power [31,36]. For the near term of 5 to 10 years, the potential is estimated at least 800 MW for mini and micro hydro, which consists of hydro power plants, including storage based and run-of-the-river plants [15].

Table 3. Biomass resource potential per each province [30].

No.	Provinces	Municipal solid waste (Tonne/year)	Electricity production potential (MWh/ year)	Animal manur (Tonne/year)	Electricity production potential (MWh/year)	Crop Residue (Tonne/year)	Electricity production potential (MWh/ year)
1	Badakshan	132086	29059	2407756	452658	143550	598604
2	Badghis	68897	15157	787318	148016	138570	577837
3	Baghlan	126100	27742	1579481	296943	345860	1442236
4	Balkh	181785	39993	1045096	196478	415330	1731926
5	Bamyan	62123	13667	761429	143149	68720	286562
6	Daykundi	64021	14085	1114886	209599	46620	194405
7	Farah	70430	15495	865514	162717	90030	375425
8	Faryab	138408	30450	971824	182703	323940	1350830
9	Ghazni	170645	37542	1212019	227860	322200	1343574
10	Ghor	95951	21109	1236102	232387	97470	406450
11	Heart	259880	57174	2258046	424513	415470	1732510
12	Helmand	128407	28250	2033377	382275	482920	2013776
13	Jawzjan	74767	16449	403238	75809	217440	906725
14	Kabul	576744	126833.6	367493	69089	111600	465372
15	Kandahar	168061	36973	2164179	406866	266440	1111055
16	Kapisa	61291	13484	911902	171438	106020	442103
17	Khost	79833	17563	1956218	367769	90290	376509
18	Kunar	62605	13773	1669748	313913	107680	449026
19	Kundoz	139255	30636	2066156	388437	446790	1863114
20	Laghman	61919	13622	1431920	269201	156870	654148
21	Logar	54473	11984	386770	72713	202020	842423
22	Nangarhar	209656	46124	3333441	626687	419610	1749774
23	Nimroz	22864	5030	252388	47449	71730	299114
24	Nooristan	20571	4526	832461	156503	22420	93491
25	Paktia	76650	16863	1263836	237601	147740	616076
26	Paktika	60415	13291	768982	144569	114090	475755
27	Panjshir	21331	4693	197212	37076	44560	185815
28	Parwan	92214	20287	852559	160281	135580	565369
29	Samangan	53845	11846	282627	53134	99890	416541
30	Sar-i-pul	77672	17088	664815	124985	147900	616743
31	Takhar	136320	29990	1434745	269732	365550	1524344
32	Urozgan	48691	10712	847924	159410	140190	584592
33	Wardak	82870	18231	451688	84917	127030	529715
34	Zabul	42238	9292	374493	70405	62700	261459
	Sub Total	3723015	819063	39187641	7367277	6494820	27083399
Total energy production from biomass resources					**35,289,739 (MWh/year)**		

Table 4. Hydro potential resources by zone [30].

No.	Zone	River	Potential [MW]
1	Kabul	Kabul	408
2		Panjshir	400
3		Laghman	44
4		Kunar	1089
5	Panj-Amu	Panj	9050
6		Amu	9110
7		Kokcha	1927
8		Kunduz	50
9	Northern	Jawzjan	460
10		Balkh	300
11	Harirod-Murghab	Harirod	102
12		Murghab	100
13	Helmand	Helmand	190
14		Farah Rod	80
Total			**23,310**

4.5. Geothermal energy

The geothermal resource as a clean, sustainable, cost-effective and environmentally friend energy also has got attention for electricity generation in all over the world. In Afghanistan, Historically, it has been only used for therapeutic bathing. This application is still one of the important utilization of geothermal energy in Afghanistan. The active geothermal systems in the country are located in the main axis areas of the Hindu Kush, which runs along the Heart fault system, up to the Wakhan corridor in the Afghan Pamir [37,38]. The Herat-Panjshir east-to-west striking geo suture is a deep-seated strike-slip fault of up to 700 km into the mantle. In addition, there are similar structures along the Chaman-Moqor NE-SW striking fault system, the Sarobi-Altimore Northeast-to-Southwest arcuate fault system and other secondary faults which cover most of the regions of the country [33].

The geothermal energy in Afghanistan can provide an abundant high-temperature water or steam used to produce electricity and supply part of the electricity demand. According to Saba et al. (2004), the average geothermal power plant requires a total of only 400 square meters of land to produce a gigawatt of power over a period of 30 years, which is incomparable to the huge acreages needed for other power plant developments. Currently, there is not any power production or project of geothermal energy throughout the country. However, totally 70 sites have been identified and about 5-20 MW capacity power plant can be built at each [15]. The major areas of the country for the available geothermal reserves are depicted in Figure 11 [39].

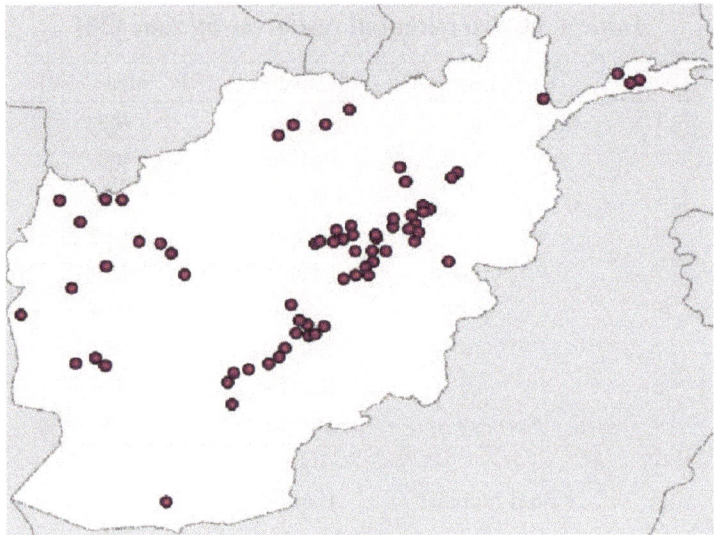

Figure 11. Major locations of geothermal resources [39].

5. Recommendations

The power sector of Afghanistan is still heavily dependent on non-renewable sources and unreliable and unstable power imports. The utilization of renewable energy sources is an inevitable challenge for the country. The ANREP set the target to 10% of the total demand (350–450 MW) by 2032. Although, this target is still small compared to the total demand (3502 MW) of the same year, there are some barriers and challenges that hamper the development of renewable energy in the country. The general instability, lack of powerful policy, high cost of RE, lack of information and public awareness programs on the benefits of RE, lack of skilled personnel and lack of private sector participation are the key obstacles that prohibit the RE implementation.

Therefore, the government is responsible to provide a productive environment and enough motivation and incentives among the various energy users and sector and present adequate demonstration and justification of renewable energy technologies. In addition, there is lack of general information and incentives on the technology, characteristics and benefits of RET. Primary information sources such as social and electronic media, institutions, effective seminars and other useful activities could play a significant role for public awareness which are not possible without the help of government agencies and other energy involved organizations in the country. On the other hand, Renewable energy sources are not commercially competitive with their fossil fuel alternatives. Hence, government support is necessary in order to accelerate the commercialization of renewable energy. This can be done through provision of adequate fiscal and financial incentives.

It can be concluded from renewable energy policy that the Government alone cannot address a massive electrification and there is a need to involve the private sector and create public private partnerships in generation and service delivery. Government must work together with energy involved and private organizations and international donor agencies for providing economic support for renewable energy development. Besides, there is an urgent need for technical assistance programs designed to increase the planning skills and understanding of renewable energy technologies by utilities, regulators and other institutions involved.

Finally, as the energy demand is continuously rising and vice versa the existed power stations in the country are out of their useful life period and mostly older than 40 years and need to be rehabilitated. Besides, comparing the tremendous potential available in the country, the ANREP should be revised and increase the target of generation from RE to supplement the future demand and fulfil the socio-economic development, energy security and sustainability in Afghanistan.

6. Discussion and Conclusion

This paper reviews the potential of renewable energy especially solar, wind, hydro, biomass and geothermal in Afghanistan. From these renewables, the available resource potential data of geothermal energy is limited in the country. On the other hand, it is included in the country's rural renewable energy policy that, efforts will be undertaken to demonstrate the potential of geothermal energy as a pilot project in the feasible areas and also a database of geothermal sites will be developed.

However, from the continuing discussion, it is concluded that the total theoretical potential of renewable energy excluding geothermal energy is about 66000 GW, and total feasible potential is 297 GW, where the major portion dominates by solar energy. As a comparison, the total feasible potential of renewable energy is almost 66 times the total demand of 2032. As a conclusion, if only 2% of total feasible potential could be utilized, it can lonely satisfy the total energy demand of the country. Besides, as most of the country's population live in the rural area with 36% under the absolute poverty line [6], and also the expansion of transmission and distribution networks are quite costly. Thus, the utilization and development of these resources which are dispersed throughout the country, will stimulate socio-economic growth and will decrease poverty and the stage of unemployment people by creating various jobs in the country. Moreover, due to a good strategic location of Afghanistan, it will create an opportunity to export the excessive power and expand the trade with other countries. Finally, higher penetration of renewable energy means increasing energy security and reducing dependence on unreliable and unstable imported power from the neighboring countries.

Acknowledgments

The authors appreciate the support of University of the Ryukyus, Kabul Polytechnic University, and Renewable Energy Department of Afghanistan for this research through the JICA (PEACE project) in 2016.

Conflict of Interest

The authors declare there is no conflict of interest.

References

1. Ershad AM (2014) potential of solar photovoltaic and wind power plants in meeting electricity demand in Afghanistan," Renewable and clean energy: [dissertation].Ohio: Dayton Univ.

2. The World Bank. Working for a world free of poverty. Available from: http://data.worldbank.org/indicator/SP.POP.TOTL.

3. ADB (2014) Technical Assistance Report, Islamic Republic of Afghanistan. Renewable Energy Development: Capacity Development Technical Assistance (CDTA), Project Number: 47266-001.

4. Ghalib A (2015) Ministry of energy and water: pp tx RED ppp To MEW.

5. Ministry of Energy and Water (2014) Afghanistan National Renewable Energy Policy. Final Draft.

6. Mundi index. Available from: http://www.indexmundi.com/map.

7. Kampan PA (2015) Balancing Political, Economic, and Scientific Objectives. Tanielian, Energy Reform in ASEAN. *Int J Emerg Electric Power Syst* 16: 297-311.

8. Hadjipaschalis I, Christou C, Poullikkas A (2008) Assessment of Future Sustainable Power Technologies with Carbon Capture and Storage. *Int J Emerg Electric Power Syst* 9.

9. John I, Peter M (2012) Afghanistan Resource Corridor Development: Power Sector Analysis, Australian AID.

10. Alamyar KM (2014) Renewable Energy for sustainable Development.

11. ADB (2013) Power sector master plan, Technical Assistance Consultants Report. Islamic Republic of Afghanistan: Project Number 43497.

12. Samadi AR (2013) Afghanistan Strategies on Renewable Energy. DABSCEO Presentation.

13. Nasrati AA (2015) Sustainable Energy for All Afghanistan. Country Presentation: SE4ALL Consultation Workshop, Manila, Philippines.

14. Bhandari NM, Burt G, Dahal KP, et al. (2007) Dispatch Optimization of Renewable Energy Generation Participating in a Liberalized Electricity Market. *Int J Emerg Electric Power Syst* 8.

15. Ministry of Energy & Water, Ministry of Rural Rehabilitation & Development. (2013) Afghanistan Rural Renewable Energy Strategy. Islamic Republic of Afghanistan: Action Plan until 2018; Draft version 1.

16. Shoaib A, Ariaratnam S (2016) A Study of Socioeconomic Impacts of Renewable Energy Projects in Afghanistan. *ICSDEC* 145: 995-1003.

17. Farooq MK, Kumar S (2013) An assessment of renewable energy potential for electricity generation in Pakistan. *Renew Sust Energ Rev* 20: 240-254.

18. Ershad AM, Robert JB, Kevin H (2015) Analysis of solar photovoltaic and wind power potential in Afghanistan. *Renew Energ* 85:445-453.

19. Rezaei M, Chaharsooghi SK, Abbaszadeh P (2013) The Role of Renewable Energies in Sustainable Development: Case Study Iran. *Iranica J Energ Environ* 4: 320-329.

20. United Nations Economic Commission for Europe. (2013) Assessment On Clean Infrastructure Development In Turkmenistan. Available from: https://www.unece.org/fileadmin/DAM/ceci/doc uments/UNDA_project/PPP_Assessment_Turkmenistan.pdf

21. Nachmany M, Fankhauser S, Davidová J, et al. (2015) Climate Change Legislation in Tajikstan, The 2015 Global Climate Legislation Study A Review of Climate Change Legislation in 99 Countries. Available from: http://www.lse.ac.uk/GranthamInstitute/wp-content/uploads/2015/05/TAJIKISTAN.pdf

22. In-Depth Energy Efficiency Review Tajikistan (2013) energy charter secretatiat. Available from: http://www.energycharter.org/fileadmin/DocumentsMedia/IDEER/IDEER-Tajikistan_2013_en.pdf

23. Halder PK, Paul N, Joardder MUH, et al. (2015) Energy scarcity and potential of renewable energy in Bangladesh. *Renew Sust Energ Rev* 51: 1636-1649.

24. Khoury J, Mbayed R, Salloumb G, et al. (2016) Review on the integration of photovoltaic renewable energy in developing countries—Special attention to the Lebanese case. *Renew Sust Energ Rev* 57: 562-575.

25. Daniel K, Goran S (2004) Fundamentals of Power System Economics, UK: John Wiley & Sons Ltd, ISBN: 0470845724.

26. Da Afghanistan Breshna Sherkat (2015) Energy Supply Improvement Invest-ment Program, Afghanistan, Environmental Assessment and Review Framework.

27. John I, Peter M (2012) Afghanistan Resource Corridor Development, Power Sector Analysis, Australian AID.

28. Ashna GF (2014) Afghanistan Energy Report, Da Afghanistan Brishna Shirkat, IEEJ. Available from: https://eneken.ieej.or.jp/data/5585.pdf.

29. Farooq MK, Kumar S (2013) An assessment of renewable energy potential forelectricity generation in Pakistan. *Renew Sust Energ Rev* 20: 240254.

30. Ministry of energy and water (2015) Renewable energy department database. Islamic Republic of Afghanistan. Available from: http://www.red mew.gov.af/database/ren-database/.

31. Ministry of Energy & Water, Ministry of Mines, Ministry of Rural Rehabilitation & Development (2013) Energy Sector Strategy, Islamic Republic of Afghanistan: Afghanistan national development strategies.

32. NREL. Afghanistan Resource Maps and Toolkit, US: Leading clean energy innovation. Available from: http://www.nrel.gov/international/ra afghanistan.html.

33. Zaher M. Afghanistan Initial National Communication To the United Nations Framework Convention on Climate Change, Islamic Republic of Afghanistan: National environmental protection agency.

34. Bansal RC, Zobaa AF, Saket RK (2005) Some Issues Related to Power Generation Using Wind Energy Conversion Systems. An Overview. *Int J Emerg Electric Power Syst* 3.

35. Milbrandt A, Overend R (2011) Assessment of Biomass Resources in Afghanistan, US department of energy: NREL/TP-6A20-49358 Technical Report, Contract No. DE-AC36-08GO28308.

36. Meisen P, Azizy P (2008) Rural Electrification in Afghanistan, How do we electrify the villages of Afghanistan? Global Energy Network Institute.

37. Saba DS, Najaf ME, Musazai AM, et al. (2004) Geothermal Energy in Afghanistan, Prospects and Potential. SABA et al 1feb04. New York, USA. & Afghanistan Center for Policy and Development Studies, New York University Kabul, Afghanistan.

38. Ministry of energy & water, Ministry of rural rehabilitation & development (2013) Afghanistan Rural Renewable Energy Policy, Islamic Republic of Afghanistan, Final Draft.

39. Brown WT, John V, Tarek A, et al. (2011) Feasibility of Renewable Energy Technology at the Afghanistan National Security University, A Site Specific Study Focused on Potential Renewable Energy Technologies in Northwest Kabul, Afghanistan. ERDC/CERL TR-11-12.

Evaluating the potential of renewable diesel production from algae cultured on wastewater: techno-economic analysis and life cycle assessment

Ankita Juneja [1,2,*] **and Ganti S. Murthy** [1]

[1] Biological and Ecological Engineering, Oregon State University, Corvallis OR 97331, USA
[2] Agricultural and Biological Engineering, University of Illinois Urbana Champaign, Urbana IL 61801, USA

* **Correspondence:** Email: ankitajuneja27@gmail.com

Abstract: Algae, a renewable energy source, has an added advantage of consuming nutrients from wastewater and consequently aiding in wastewater treatment. The algae thus produced can be processed using alternative paths for conversion to fuels. However, due to high moisture content of algae, wet algae processing methods are being encouraged to avoid the dewatering cost and energy. Hydrothermal liquefaction is one such technology that converts the algae into high heating value bio-oil under high temperature and pressure. This bio-oil can be further upgraded to renewable diesel (RD) which can be used in diesel powered vehicles without any modifications. The objective of this study is to evaluate the economic viability and to estimate the energy use and greenhouse gas (GHG) emissions during life cycle of RD production from algae grown in wastewater using hydrothermal liquefaction. Economic analysis of RD production on commercial scale was performed using engineering process model of RD production plant with processing capacity of 60 Mgal wastewater/day, simulated in SuperPro designer. RD yields for algae were estimated as 10.18 MML/year with unit price of production as $1.75/RD. The GHG emissions during life cycle of RD production were found to be 6.2 times less than those produced for conventional diesel. Sensitivity analysis indicated a potential to reduce ethanol production cost either by using high lipid algae or increasing the plant size. The integrated economic and ecological assessment analyses are helpful in determining long-term sustainability of a product and can be used to drive energy policies in an environmentally sustainable direction.

Keywords: microalgae; hydrothermal liquefaction; renewable diesel; techno-economic analysis; life cycle assessment

---

1. Introduction

Total energy consumption of United States in 2013 was 1.04×10^5 quadrillion joules, of which about 82% comes from fossil reserves and only 9.3% is contributed by renewable energy. Transportation sector is one of the biggest consumers of energy, which accounts for nearly one third of the total energy consumption worldwide and contributes to about 21% of greenhouse gas (GHG) emissions [1]. Concerns of depleting fossil fuels and increasing environmental burden has encouraged the exploration of additional renewable sources. Interest in algal biofuels can be attributed to the possibility of year round production at higher productivities compared to terrestrial crops, non-competition with food crops, reduced need for arable land and water treatment benefits with nutrient cycling. It has been studied that algae can produce 10–50 times more oil per unit area than conventional oil seed crops such as canola, jatropha and oil palm [2]. Another considerable advantage of algae is its ability to consume nutrients from wastewater and thus help in wastewater treatment, which solves another big challenge of clean water requirement. One of the challenges for microalgal derived biofuels is the dewatering of algae, as the alternative routes of lipid extraction with hexane, pyrolysis or gasification require the use of dry algae (85–95% solids) and drying is one of the most cost intensive process in the whole route of oil production, which can add up to 30% of the total cost [3]. Efforts are required to use the wet biomass directly for biofuel production to eliminate the energy required for dewatering/drying. Although wet lipid extraction methods including ultrasound-assisted extraction [4], simultaneous distillation and extraction process [5], microwave-assisted extraction [6] and supercritical fluid extraction [7] have been investigated by various researchers, the technologies require high cost and energy input. Hydrothermal liquefaction (HTL) is one of the technologies that fits this criteria of using wet algae. HTL is a thermal process of converting algae into high heating value bio-oil under high temperature and pressure.

1.1. Hydrothermal liquefaction

Hydrothermal liquefaction (HTL) is gaining attraction as an alternative route for biofuels production from algal biomass. HTL reactions, involving dehydration, deoxygenation, and decarboxylation [8], occur at elevated temperatures (250–380 °C), high pressure (5–30 MPa) and varying residence times (3–60 min) [9,10]. At such high temperatures and pressure, water still stays in liquid phase but the dielectric constant and the density of water is decreased relative to water at normal temperature and pressure, which causes water to become non-polar, highly reactive and miscible for organic components; thereby working as a catalyst. This results in hydrocarbon becoming more soluble in water [11]. HTL uses wet algae without the need for cell disruption, which eliminates the drying costs. Another advantage of HTL over traditional lipid extraction is the possibility of using low lipid algae as feedstock which is a significant benefit in case of fast growing algae with high carbohydrates and proteins but low lipid productivity, such as those commonly found in waste water treatment plants.

Biocrude obtained from HTL must be hydrotreated for reducing the overall oxygen content to produce usable form of fuel, renewable diesel. Renewable diesel yield from HTL (104,000 m^3/y) was observed to be 12% better than that of biodiesel from lipid extraction (91,300 m^3/y) [12]. Renewable diesel (1) can be directly used in diesel-powered vehicles without modifications, (2) is compatible with current diesel distribution infrastructure, (3) can be produced using existing oil refinery capacity, (4) can be used in advance emission control devices due to ultra-low sulfur content and (5) exhibits better performance than diesel [13].

1.2. Techno-economic analysis and life cycle assessment

Algal biofuels are currently proven on lab scale but commercialization of the process is still in its infancy. Long term viability of any feedstock or process not only depends on the yield of the final product, but also on sustained production capacity, maximized profits and minimized environmental burden. It is important to determine the overall energy efficiency along with capital and operating cost estimations to investigate the sustainability of a process. Techno economic analysis is an important tool to analyze the cost and energy viability of the process on large scale. Environmental impact, in terms of GHG emissions and fossil energy use, of the process can be calculated by performing the life cycle assessment. To obtain better insight into the current state of biofuel technology, this study evaluates the overall economics of renewable diesel production from algal biomass. The study will also help in identifying the key parameters/operations in the renewable diesel (RD) production process, which can be targeted for further improvements.

Life cycle analysis is a useful technique to assess impact of products, processes and services on the environment and can play an important role in comparing renewable diesel with other fuel alternatives based on environmental impact. Most of the previous studies performed on assessing environmental impacts of algal biofuels focus on biofuel production focused on production processes that involved fresh water/sea water with use of supplied nutrients [14,15]. This study, however, assess the production models that use wastewater as a source of water and nutrients to achieve a dual goal of biofuel production and wastewater treatment. Therefore, the overall objective of this study is to perform a comprehensive techno-economic analysis and limited life cycle assessment of RD production to analyze the economics and environmental impact of the production process and identify the key process that have the largest contribution in the overall RD production process.

2.   Methods

2.1. Algae strain

Chlorella vulgaris (*C. vulgaris*) has shown a high potential for wastewater treatment during its growth [16]. *Chlorella vulgaris* is one of the promising algae strains for biofuels production, with high productivity (1.06 g/L/day), high rate of CO_2 fixation (1.99 g/L/day), and tolerance to high concentrations to CO_2 and compounds such as nitrogen oxides (NOx) and sulfur oxides (SOx) [17]. These characteristics make these algae suitable for growth in open ponds using wastewater as a nutrient source with added benefits of flue gas utilization to meet algae carbon requirements. Therefore *C. vulgaris* was used as algal feedstock to develop the model, as high growth, low lipid content algae such as *C. vulgaris* is suitable for hydrothermal liquefaction (HTL). Lipids, carbohydrates and protein content in *C. vulgaris* in this study is assumed to be 25%, 9% and 55% respectively on dry basis [18].

2.2. Techno-economic analysis

A process model was developed for treating incoming wastewater of 227 million L/day (60 million gallon/day) from local community using Super Pro Designer (Intelligen, Inc., Scotch Plains, NJ). The process model contains five sections: growth, harvesting, hydrothermal liquefaction, bio-oil hydrotreating and co-product recovery and utilization (catalytic hydrothermal gasification) (Figure 1).

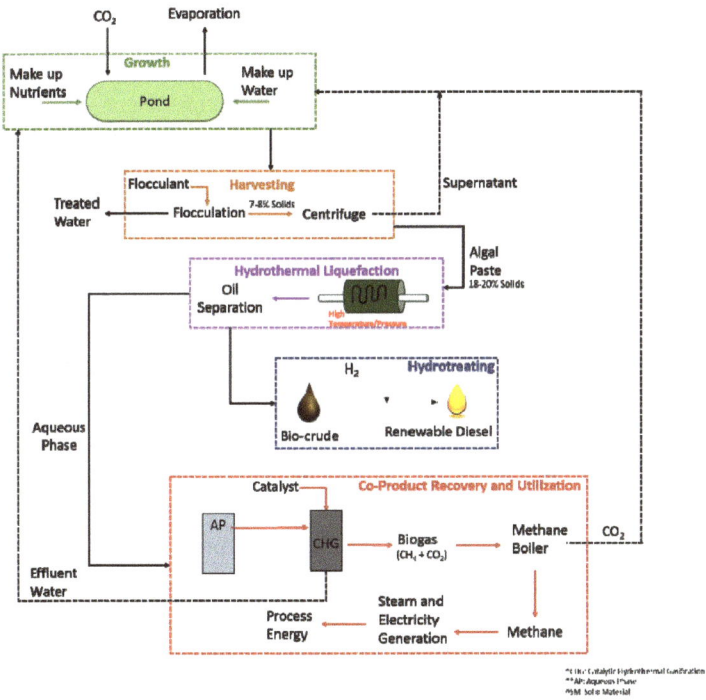

Figure 1. Schematic illustration of modeled bio-oil plant.

2.3. Technical assumptions

It was assumed that algae production facility receives adequate solar radiation to obtain productivity of 25 g/m^2/day, and plant operates throughout the year for 330 days. The concentration of algae going out the pond was 0.38 g/L (4 days retention time). Productivity of 25 g/m^2/day and a need for treatment of 227 million L/day wastewater, requires construction of 85 ponds of 4 ha each with a total area of 340 ha. The total algae flow from the ponds to the harvesting section was 3550 kg/h. Accounting for a 2% loss of algae during flocculation and 0.01% during centrifugation, a total of 3478 kg/hr algae was processed for hydrothermal liquefaction, which produced 1222 kg/hr biocrude oil. Hydrotreating of biocrude produced 990 kg/hr (340 gal/hr) RD. The remnants of HTL (16,455 kg/hr, which includes ~ 4% organic matter and ~ 95% water) were used to produce methane rich biogas (60% methane and 40% CO$_2$) by catalytic hydrothermal liquefaction, producing

710 kg/hr biogas. The CHP (combined heat and power generation) unit with efficiency of 38% electricity production [19] using direct biogas produced 2412 kW electricity, most of which was used to fulfil the plant requirements and the rest was exported out of the plant.

2.4. Economic assumptions

The plant is assumed to operate for 330 days/year. Price basis for all the economic calculations was taken as 2015. Costs of specific equipment (open ponds, clarifier, centrifuge, hydrothermal liquefaction reactor, hydrothermal gasification reactor and hydrotreating reactor) for renewable diesel production process were calculated based on previously reported cost models [20,21,22] (listed in supplementary Table 1). Costs were adjusted to year 2015 using the equation 1:

$$Cost\ in\ 2015 = Cost\ in\ baseyear * \left(\frac{2015\ index}{base\ year\ index}\right) \tag{1}$$

Cost of other equipment was based on the built-in cost models in SuperPro designer. The scale up of the equipment was done using the conversion from equation 2:

$$Scaled\ cost = original\ scale\ cost * \left(\frac{Scale\ up\ capacity}{Original\ capacity}\right)^n \tag{2}$$

where n is the scaling factor (typically 0.6–0.7).

2.5. Algae cultivation

Algae plant was assumed to be based on a modular design based on raceway ponds (L:W:D = 890 m :44 m :0.3 m). The growth rate of algae was calculated based on Monod kinetic parameters of algal growth with nitrate as limiting substrate calculated by laboratory experiments (equation 3).

$$\mu = \mu_{max} \frac{S}{S+k_s} \tag{3}$$

where μ is the specific growth rate of microalgae, μ_{max} is the maximum specific growth rate, S in this study is the concentration of nitrate in wastewater and k_s is the half saturation constant.

Mixing of algae slurry in the open pond using paddle wheels is necessary to avoid sedimentation, distribute nutrients, and maintain a uniform temperature and pH to enhance algae productivity [23]. The power requirement for the paddle wheel is calculated using equation 4 [20].

$$P = \frac{QW\Delta d}{102e} \tag{4}$$

where P is the power requirement (kW), Q is the flow of the culture (m³/s) (calculated by velocity × cross-sectional area), W is the specific weight of the culture (kg m⁻³), Δd is the head loss of water, e is the efficiency of paddle wheel (40% assumed [22]) and 102 is the conversion factor. The velocity of water flowing with paddle wheel is assumed to be 15 cm/s [20] and the cross-sectional area can be calculated as depth × width/2 (6 m²). Additional head loss of water in the pond (Δd) occurs along the length of the pond and as water flows around two 180° bend with baffles. Both the head losses can be calculated using the Manning's equation. Total head loss can be calculated as the summation of the two losses (equation 5) [22].

$$\Delta d = v^2 n^2 \left(\frac{L}{R^{\frac{4}{3}}}\right) + \frac{Kv^2}{2g} \tag{5}$$

where v is the mean velocity (m/s), n is the roughness factor (0.18 for clay lining), R is the channel hydraulic radius (m), L is the channel length (m), K is the kinetic loss coefficient for 180° bends (theoretically 2).

With surface area (cross-sectional area) of 6 m^2 and wetted perimeter as 20.6 m, the hydraulic radius is calculated as 0.29 m. Total head loss was calculated as 0.0238 m (0.0046 m around two 180° bends and 0.0192 m along the total channel length of 1790 m).

The pond is assumed to be built with compacted earth and lined with clay locally available avoiding the high cost of plastic liner [22]. A report suggested an increase in pond construction from $136,000 for clay lined to $277,000 with plastic lining (year 2010 reference) for a 4-ha raceway pond [22]. The total cost of one open pond was $154,600, and the breakdown of the cost is presented in Table 1. Flue gas (rich in CO_2) from a power plant (Table 2) at a distance of up to 4.5 km (three miles) is transported by pipes for addition to the growth pond. It is assumed that utilization efficiency of CO_2 in flue gas directly injected into algae ponds is more than 90% [24]. Some water is recycled back from centrifugation step (along with some nutrients), and rest of makeup water is provided by wastewater stream assumed to be transported from treatment plant 2 miles away. The composition of wastewater is presented in Table 2 [25]. Algae can utilize nutrients available in wastewater for growth which also helps in treatment of wastewater. Wastewater stream, thus, reduces the use of makeup nutrients required for algal growth. Any deficient nutrient was replenished with pure nutrient (considering only nitrates and phosphates in this study).

Table 1. Breakdown of the open pond cost [47,51].

	Cost (per hectare) (Cost updated to 2015)	Cost of 1 open pond (4 ha) (Cost updated to 2015)
Cost of open pond site	$3640/ha	$14560
Cost of open pond construction (wall and structure)	$14450/ha	$57800
Paddle wheel	$7285/ha	$29140
CO_2 delivery (pipes and pumps)	$9180/ha	$36720
Internal distribution of CO_2 (sumps)	$4090/ha	$16360
Total open pond cost	$38645/ha	$154600

2.6. Algae harvesting

It is assumed that algae grows to a steady-state concentration and is continuously harvested at the rate equal to the growth rate. Algae is harvested in two steps: bulk dewatering and thickening. Dewatering can be achieved using flocculation or floatation. Although floatation is an effective process as algae floats up faster than settling down in flocculation [3], power consumption in pumping air for dispersed air flotation if high, making the process expensive. Also, the choice of coagulant/collector is very critical in air flotation [26]. Therefore, flocculation process was chosen for bulk dewatering step using cationic starch as flocculent [27]. A high sedimentation of 95% algae is reported to be achieved with 30 minutes residence time with a concentration of 10 mg flocculent per liter of culture [27]. This step yields a concentration of 15 g/L. Each clarifier used (settling pond) has a working volume of 12,500 m^3. A total of 3.75 ha of land was used for three settling units required for the operation. The thickening of algae is obtained by centrifugation process, providing the final concentration of 200 g/L (20% solids) algae. The throughput of centrifuge was 90 m^3/h with the cost of one unit as $277,000. In the current study, use of centrifuge was assumed as final step in algae harvesting process, however other studies have used sun drying, belt press, bed drying as other options to increase algal concentration [12].

Table 2. Composition of flue gas and wastewater used in algal growth.

Composition of flue gas [52]		Composition of wastewater [25]	
Component	*Mass composition*	*Component*	*Mass composition*
Nitrogen	76.5%	Water	99.7%
Carbon dioxide	12.8%	Other Solids	0.12%
Water	6.2%	Nitrates	800 ppm
Oxygen	4.4%	Inorganic carbon	400 ppm
NOx	420 ppm	Salts	100 ppm
SOx	420 ppm	Sulfates	50 ppm
Carbon monoxide	50 ppm	Phosphates	15 ppm

2.7. Bio-crude production

The solids concentration reported in literature for HTL of algae is 10–20% [28,29]. For the purpose of this study, it was assumed that the 20% solids concentrated algae obtained from previous centrifugation step, transported using positive displacement pump, which are offered by vendors to pump slurries with 20% solids [30], is directly used in HTL reactor. The optimal HTL process conditions for maximum oil productivity (~ 37% reported) [28] reported in literature [10,28] were chosen for the process modeling 340 °C, 200 bar pressure and residence time of three minutes. Choice of residence time was based on studies that have indicated possibility of higher yields with shorter residence times (1–10 minutes) [31,32]. The bio-crude yield was calculated based on the additive relation proposed by Biller and Ross [18], with protein yield as 10% [11], carbohydrate yield as 32% and lipid yield as 95% [33]. For algae flow rate of 3478 kh/hr, the biocrude yield was calculated as 35% (1221.9 kg/hr). The pressurized slurry is preheated at 335 °C before entering into the reactor. The reactor was chosen to be plug flow reactor because of lower economics at this scale (capital and operating cost) than continuously stirred tank reactor, used in other studies [8]. The HTL conditions and elemental analysis of bio-oil are reported in Table 3.

Table 3. Conditions of hydrothermal liquefaction and properties of bio-oil [21].

Temperature (°C)	340 °C
Pressure	200 bar
Solids concentration in feed (wt.%)	20%
Residence time	3 min
HTL oil yield (wt.%)	37%
Aqueous co-product (wt.%)	36%
HTL bio-oil analysis wt.%	
C	77.3%
H	9.4%
O	6.1%
N	6.2%
S	0.65%
Bio-oil moisture content	2.8%
Bio-oil HHV	24 MJ/kg

2.8. Hydrothermal gasification

The coproduct stream from algal oil extraction via hexane extraction or HTL is rich in organic matter, which makes it suitable to be converted to biogas [8,15]. The most common route of conversion is anaerobic digestion, but in case of HTL, aqueous byproducts are rich in nitrogen (about 70% of feedstock nitrogen ends up in aqueous phase). The ideal range of C:N for successful anaerobic digestion is 20:1–30:1. Jena et al. [34] measured 3.92% carbon and 1.84% nitrogen in the aqueous phase after HTL of *Spirulina platensis*, indicating a C:N ratio of 2.1:1. High nitrogen content in the aqueous phase makes it unsuitable to be digested in anaerobic digester. Such sludge is suitable for treatment in catalytic hydrothermal gasification (CHG), a process that converts the nitrogen and carbon of the wet organic matter to biogas with use of a catalyst. The CHG treatment is very efficient for the conversion of organic carbon in the aqueous phase (99.2%) [35]. Gasification dominates in temperatures above 647 K [35]. CHG leads to formation of a methane rich (60%) biogas [36] and the conversion of organics is independent of the operating pressure after the temperatures are above 647 K [37].

2.9. Hydrotreating of HTL biocrude

The biocrude obtained after hydrothermal liquefaction has to be upgraded to remove/reduce oxygen, nitrogen and sulfur before use in transportation vehicles. Hydrotraeting is done by treating the bio-oil with hydrogen (3.75% of wet bio-oil) at 350–400 °C at 3.4–10.3 MPa (300 °C and 3.5 MPa used in this study) in presence of a catalyst (CoMo/alumina-F in this study). This treatment converts the oxygen in the biocrude to CO_2 and H_2O, nitrogen to ammonia and sulfur to H_2S resulting in renewable diesel product with a density of 770 g/L [21] and a heating value of 44 MJ/kg . The cost of hydrogen production in petroleum refinery is $1.21/kg [38] and the requirement of hydrogen is 0.0375 kg/kg wet bio-oil [21].

2.10.   *Life cycle analysis*

2.10.1.   Goal definition and scoping

The goal of the study is to investigate the GHG emissions and energy use during the life cycle of renewable diesel production from algae. The study analyzed the impact of RD production process from algae via hydrothermal liquefaction. The functional unit for the model was taken as 1000 MJ of RD energy (22.73 kg or 29.5L RD). Most of the data for the LCA model was taken from the techno-economic model described in earlier sections.

2.10.2.   Process description

The model was divided into four sections: algal production and harvesting, bio-oil production and co-product recovery and bio-oil distribution. Selection of system boundary is the most critical decision which affects the computations of energy consumption and emissions. The system boundary selected for this study is shown in Figure 2.

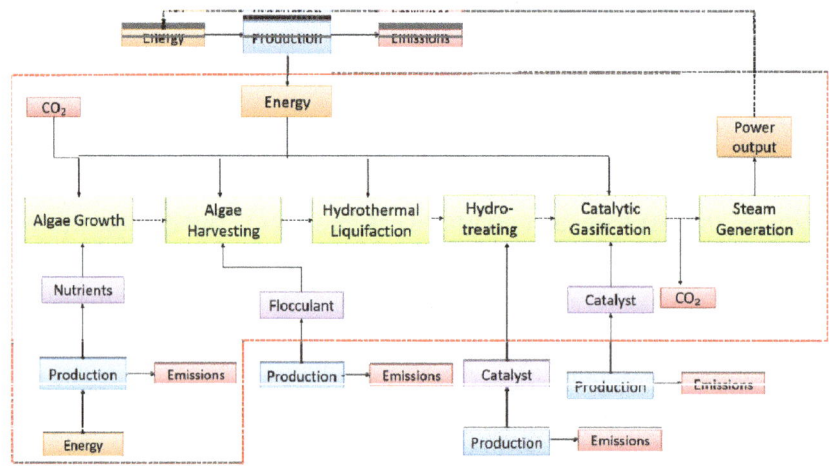

Figure 2. System boundary for life cycle analysis of RD production from algae (Red dotted line shows the system boundary considered in the study).

The co-product allocation method is another major selection that can significantly influence the results of life cycle analysis [39]. Electricity was the only co-product from the plant. System expansion (displacement) approach [39] was used to calculate the co-product credits, which assumes that the electricity produced from biogas replaces the electricity requirement for the plant operations and offsets the energy use and GHG emissions to produce the electricity that would otherwise be produced using fossil fuels. The emissions were calculated in terms of gram CO_2 equivalent using global warming potential factors of 1, 25 and 298 for CO_2, CH_4 and N_2O, respectively.

Net energy value (NEV) and net energy ratio (NER) are the two terms defined on LCA to assess the energy efficiencies and fossil fuel displacement values [40]. NEV and NER were calculated using equations 6 and 7. The NER of a system is defined as the ratio of total energy produced over the energy required for all plant operations. For a product to be sustainable (energy in the fuel > fossil energy used), NER should be greater than 1.

$$NEV = Energy\ of\ functional\ unit - energy\ use\ produce\ functional\ unit \qquad (6)$$

$$NER = \frac{\sum Energy\ produced}{\sum Energy\ requirements} \qquad (7)$$

The model was developed in GREET (Greenhouse Gases, Regulated Emissions, and Energy Use in Transportation). The values for nutrients, catalyst, hydrogen and flocculant used and energy requirement were derived from the techno-economic model.

3.   Results and Discussion

3.1. Economics

Model of bio-oil production from algae (60 million gallon wastewater per day) was simulated in SuperPro Designer (Figure 3) for a plant life of 20 years. Techno-economic model for the production of renewable diesel in the nth plant design resulted in the cost of production of RD as $1.75/L ($6.62/gal). The capital cost for bio-oil production plant were estimated to be $105 MM with renewable diesel (RD) production of 10 million liters/year (2.7 million gallon/year). Overall economics of the process are presented in Table 4. Installed equipment cost (equipment and installation cost) accounted for nearly 38% of the total direct fixed capital costs. The rest would include piping, building and other indirect costs (engineering, construction, contractor's fee). Breakdown of operating cost as facility dependent costs, raw material costs and other costs (labor and utilities) for the RD production processes is illustrated in Figure 4a. The major portion was facility dependent cost (64%), which includes the costs associated with equipment maintenance, depreciation of the fixed capital cost, and miscellaneous costs such as insurance, local (property) taxes and possibly other overhead-type of factory expenses. The raw material cost accounted for 13% of the total operating cost, with flocculant (starch) and catalyst for hydrothermal gasification being the major fraction (31% and 23% respectively). Most of the major nutrients required for growth were supplied by the wastewater (78% nitrates and 98% phosphates). Similarly, about 21% CO_2 was supplemented with the flue gas.

Table 4. Overall economics of the production of RD from algae with processing capacity of 2.5Mgal/h.

Total Investment (MM $)	104.96
Operating Cost (MM $/yr)	17.88
RD (MMgal/yr)	2.69
RD unit cost ($/l($/gal))	1.75(6.62)
Direct fixed capital cost (MM $)*	99.39
Equipment cost (MM $)	26.86
Installation cost (MM $)	10.43

*Direct fixed capital cost includes equipment and installation cost along with indirect costs.

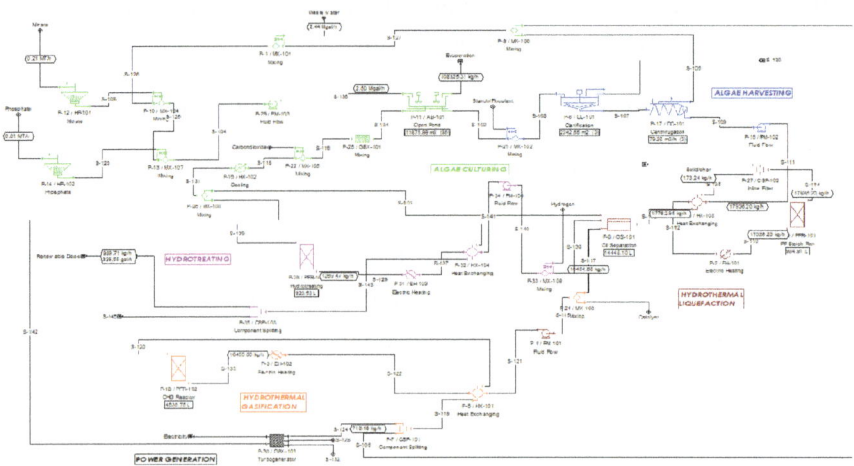

Figure 3. SuperPro model of bio-oil plant from algae.

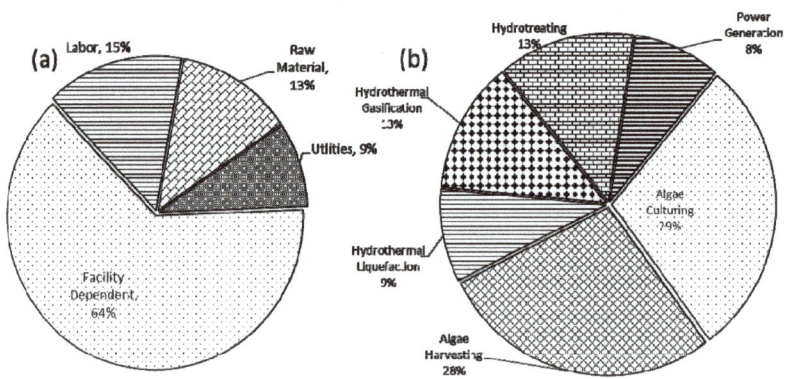

Figure 4. (a) Contribution of facility dependent, labor dependent, raw material and utilities cost during production of RD from algae and (b) distribution of RD price among different sections.

Table 5. Amount of material produced and consumed during the RD production process.

	Amount (kg/year)	Amount (kg/kg RD)	Amount (kg/functional unit)
Algae (P)	28111000	3.524	80.98
Nitrate (C)	1632000	0.208	4.70
Phosphate (C)	55000	0.007	0.16
CO_2 (C)	40411000	5.155	116.41
Starch (C)	707000	0.090	2.04
Catalyst (C)	4000	0.001	0.01
Hydrogen (C)	377000	0.048	1.09
Biocrude (P)	9677000	1.235	27.88
RD (P)	7838000	1.000	22.58

*(C) onsumption and (P) roduction.

The results of techno-economic analysis to be used for life cycle analysis are summarized in Table 5 and 6. Energy use for a section is calculated was the ratio of the energy used in that particular section to the total energy produced (energy of RD). The values of energy use for different sections of RD production system are in agreement with previously performed life cycle analysis [41].

Table 6. Energy (electricity) use in different sections of renewable diesel production.

Section	Electricity used (kWh/yr)	Electricity used (MJ/kg RD)	Energy Use (MJ/MJ RD)
Algae Growth	6508340	2.63	0.060
Harvesting	3641694	1.47	0.033
HTL	2611733	1.05	0.024
Gasification	6011259	2.42	0.055
Upgradation	2555741	1.03	0.023
Co-production	21778226	8.79	0.200
Exported	449459	0.18	0.004

3.2. Sensitivity analysis

Sensitivity analysis was performed on various inputs/operations for change in final price of renewable diesel (Figure 5). The tornado chart shows that all the parameters had significant impact on the final price of bio-oil. Lipid content of algae was the major factor affecting the price of renewable diesel, as lipids are converted to biocrude with up to 100% efficiency [33,42]. Lipid content of the strain was changed with corresponding change in protein content (nitrogen starvation leads to increase in lipids with a reduction in proteins [43]). These results are in agreement with an earlier study for production of biodiesel from microalgae that also found lipid content to be the most sensitive parameter in price of biodiesel [15]. Second major factor was the size of plant, which was shown to have considerable effect on price of RD (Figure 6). Although the first major factor is strain dependent, the second is operational parameter, which can be altered for improved economics. The changes in the amount of hydrogen for upgrading the biocrude to renewable diesel did not have a large impact on the final price because of the low quantity of hydrogen used per unit biocrude oil.

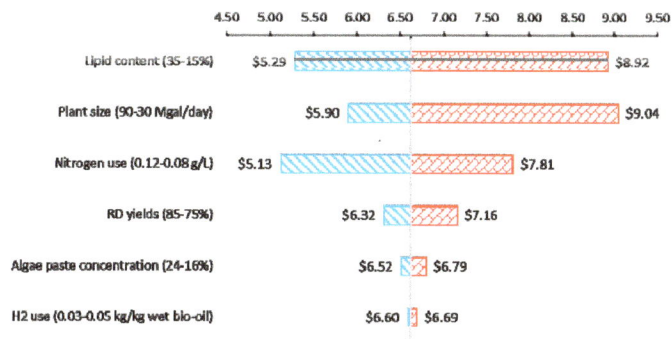

Figure 5. Effect of parameters on price of renewable diesel.

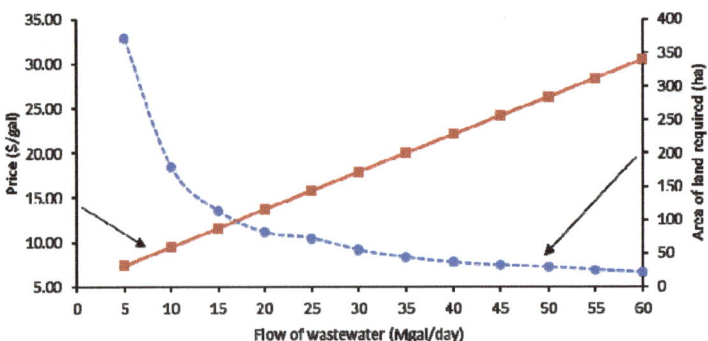

Figure 6. Effect of wastewater amount on price of RD and area required for plant.

3.3. Renewable diesel price

The price of RD was calculated as $6.62/gal, which includes for $0.86/gallon hydrotreating cost. The price of biofuel from algae has been reported to fall in the variable range of from $0.92/ gal to $42.6/gal [44]. The distribution of RD price among different sections is shown in Figure 4b. Algae culturing and harvesting contributed the largest fraction of the total cost of RD (~ 56%), in agreement with ranges reported in literature [21]. Raw material contributed to 13% of the total price of which starch added as flocculant was the highest contributor (~ 33%), followed by catalyst use (~ 24%), and hydrogen (~ 24%) and nutrients (~ 19%).

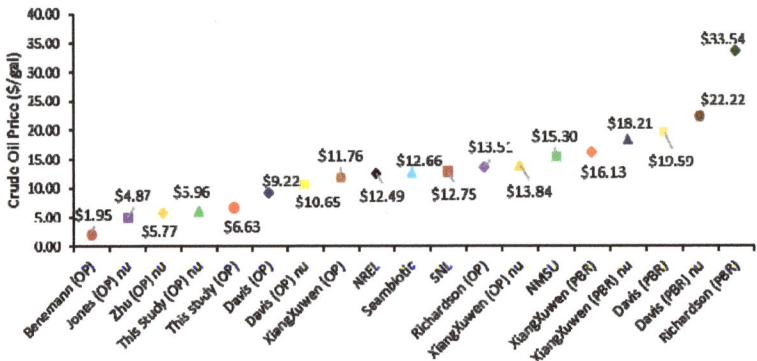

Figure 7. Comparison of biocrude prices (with biocrude and TAG prices from other studies) (Beneman [45], Davis (OP and PBR) [15], XiangXuwen (OP and PBR) [46], NREL, Seambiotic, SNL, NMSU [44], Richardson (OP and PBR) [47], Jones [21] and Zhu [8]). nu indicates not upgraded.

The comparison of the price of biocrude and renewable diesel are compared with other studies in Figure 7. All the prices were adjusted to 2015 for comparison purpose with inflation rate of 2%. In the wide range of biofuel from algae price, the price of RD in this study is on the lower limits. One of the main reasons for the low cost of production was the use of clay lining over more expensive plastic lining. The cost of RD was found to be $2.1/l ($7.93/gal) with the use of plastic lining ad increasing the cost of pond accordingly. Another reason was low makeup nutrients required for growth as wastewater adequately covered most of the nitrate and phosphate requirement. The location of plant was another favorable aspect, such that a high volume of wastewater was available for algal growth. The impact of plant size on final product price has already been demonstrated in Figure 6.

3.4. Life cycle energy use

A well to pump life cycle model for production of RD was developed in GREET. The fossil energy use for production of RD was 241.6 MJ per 1000 MJ of energy produced, which was about 80% lower than that from GREET life cycle model of conventional diesel from crude oil [48]. The fossil energy use was calculated by deducting the co-product (electricity production) from the total energy. The distribution of fossil energy use among different sections of RD production is shown in Figure 8. It can be observed that upgrading the biocrude oil had major contribution to the fossil energy use, due to hydrogen use for hydrotreating. The fossil energy use for CHP is negative implying excess electricity generation.

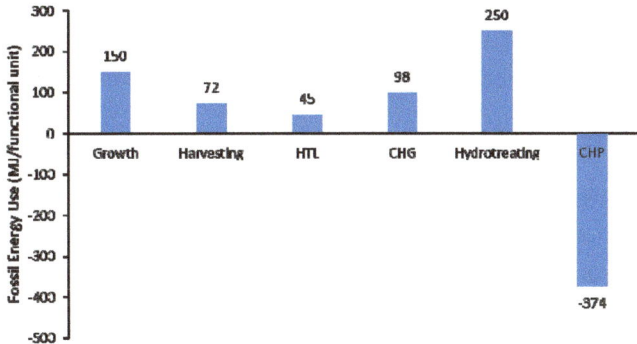

Figure 8. Distribution of fossil energy use over different sections of RD production.

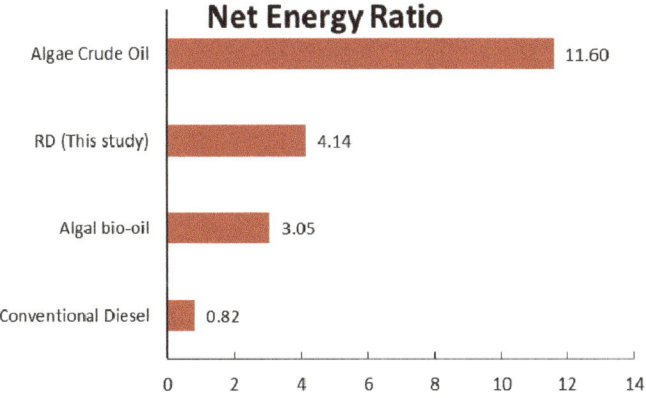

Figure 9. NER comparison for conventional diesel [48], algal bio-oil [50], RD (this study) and algal crude oil [49].

Net energy value for RD production was 758.4 MJ per 1000 MJ energy, which is higher than the conventional diesel (–207 MJ) [48] and lower than the crude oil from algae using lipid extraction (914 MJ) [49]. NER for four fuels compared is shown in Figure 9. The value of NER was less than one for gasoline indicating that the fossil energy input is higher than the energy in the fuel. The NER for algal oil is the highest, mainly because the crude oil is yet to be upgraded for use in transportation vehicles, which has a high fossil energy requirement in the form of hydrogen.

3.5. GHG emissions

The total GHG emissions during life cycle of RD production from algae were estimated to be 110 kg CO_2 equivalent per 1000 MJ of renewable diesel (–3.73 kg CO_2 eq./L RD). The GHG emissions for RD were found to be lower than the well to pump analysis of algal oil (–24 kg CO_2 equivalent per 1000 MJ) [49] and conventional diesel (21 kg CO_2 equivalent per 1000 MJ) [48]. The negative value for the GHG emission during production of RD is mainly because of system boundary selection. In well to pump analysis, the CO_2 emissions during use of fuel are not considered, which results in negative GHG emissions. Another reason for negative GHG emissions is the co-product energy available during RD production process, which displaces the GHG emissions produced by fossil fuels required to produce electricity in the plant. Similar negative values (–46.5 kg CO_2e per 1000 MJ of RD) were reported for well to pump analysis of renewable diesel production from algae earlier [41]. However, well to wheel analysis of the same study estimated GHG emission of 31 kg CO_2e per 1000 MJ of RD.

Figure 10 shows the GHG emissions in different stages of RD production. The GHG emissions during growth of algae are negative because of CO_2 sequestration in the photosynthesis process. Maximum GHG emissions were observed in hydrotreating process. The carbon flow (balance) throughout the life cycle of RD is shown in Figure 11. Red dotted line in the Figure indicates the system boundary, where sequestration of CO_2 from the atmosphere was included whereas use in the vehicle is excluded (explained in earlier sections).

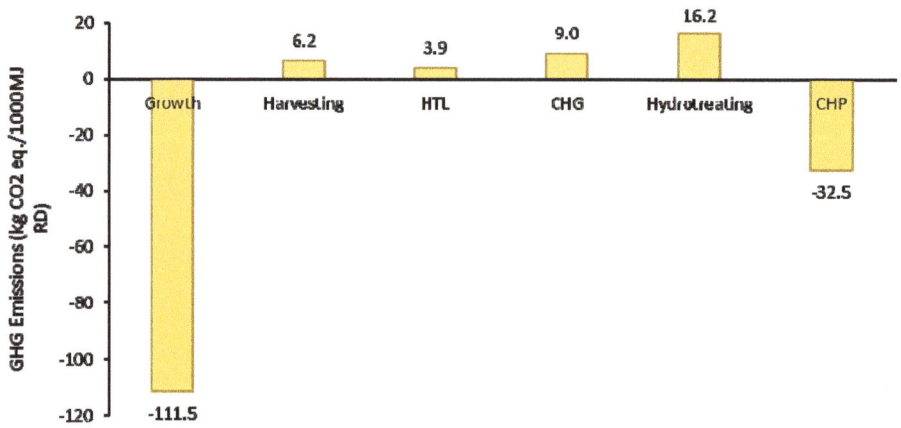

Figure 10. GHG emissions produced per functional unit during various stages of life cycle of RD production.

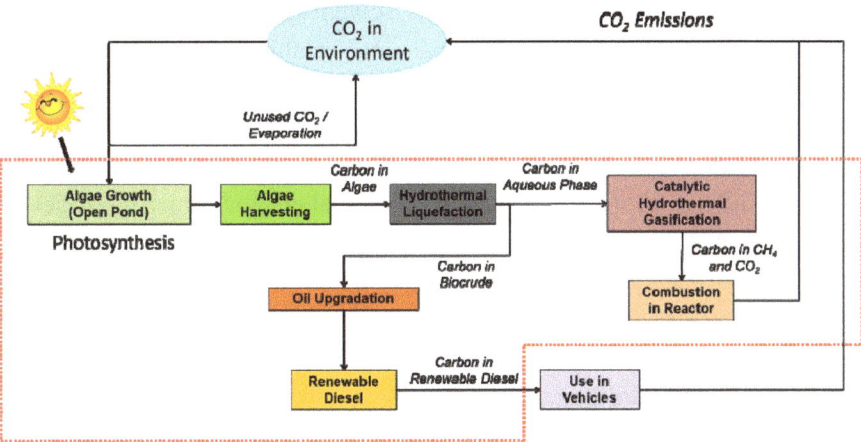

Figure 11. Carbon balance during life cycle analysis of RD production (Red dotted line indicates system boundary chosen for this study).

4.   Conclusions

A comprehensive techno-economic analysis and life cycle assessment was performed to evaluate the economic feasibility and environmental impact of the production of renewable diesel from algae via the hydrothermal liquefaction process. Cost of production of RD was calculated to be $6.62/gal, which was on the lower side of the broad range reported in literature. Highest contributors to the cost of RD were algae culturing and harvesting. The cost difference between crude bio-oil and hydrotreated renewable diesel was found to be $0.84. Lipid content of the algae and plant size were the 2 critical factors in deciding the cost of RD. The total greenhouse gas emissions in the process of RD production were –110 kg CO_2 equivalent per 1000 MJ of renewable diesel and the total fossil energy used in the process was 241.6 MJ per 1000 MJ of energy produced. The negative GHG emissions were attributed to the system boundary of the life cycle analysis. Techno-economic and life cycle assessments indicated that production of RD from algae is competitive to other alternatives to diesel production.

Conflict of Interest

All authors declare no conflicts of interest in this paper.

References

1. Ndong R, Montrejaud VM, Saint GO, et al. (2009) Life cycle assessment of biofuels from Jatropha curcas in West Africa: a field study. *GCB Bioenerg* 1: 197–210.
2. Chisti Y (2007) Biodiesel from microalgae. *Biotechnol adv* 25: 294–306.
3. Becker EW (1994) Microalgae: biotechnology and microbiology, Cambridge University Press, 10: 165.

4. Adam F, Abert VM, Peltier G, et al. (2012) "Solvent-free" ultrasound-assisted extraction of lipids from fresh microalgae cells: a green, clean and scalable process. *Bioresource Technol* 114: 457–465.

5. Tanzi CD, Vian MA, Chemat F (2013) New procedure for extraction of algal lipids from wet biomass: a green clean and scalable process. *Bioresource Technol* 134: 271–275.

6. Cheng J, Yu T, Li T, et al. (2013) Using wet microalgae for direct biodiesel production via microwave irradiation. *Bioresource Technol* 131: 531–535.

7. Halim R, Gladman B, Danquah MK, et al. (2011) Oil extraction from microalgae for biodiesel production. *Bioresource Technol* 102: 178–185.

8. Zhu Y, Biddy MJ, Jones SB, et al. (2014) Techno-economic analysis of liquid fuel production from woody biomass via hydrothermal liquefaction (HTL) and upgrading. *Appl Energ* 129: 384–394.

9. Akhtar J, Amin NAS (2011) A review on process conditions for optimum bio-oil yield in hydrothermal liquefaction of biomass. *Renew Sust Energ Rev* 15: 1615–1624.

10. Brown TM, Duan P, Savage PE (2010) Hydrothermal liquefaction and gasification of *Nannochloropsis sp. Energ Fuel* 24: 3639–3646.

11. Peterson AA, Vogel F, Lachance RP, et al. (2008) Thermochemical biofuel production in hydrothermal media: a review of sub-and supercritical water technologies. *Energ Environ Sci* 1: 32–65.

12. Delrue F, Li-Beisson Y, Setier PA, et al. (2013) Comparison of various microalgae liquid biofuel production pathways based on energetic, economic and environmental criteria. *Bioresource Technol* 136: 205–212.

13. Thomas E, David WR, Timothy RG (2010) California renewable diesel multimedia evaluation, The University of California, Davis, Berkeley.

14. Amer L, Adhikari B, Pellegrino J (2011) Technoeconomic analysis of five microalgae-to-biofuels processes of varying complexity. *Bioresource Technol* 2011: 9350–9359.

15. Davis R, Aden A, Pienkos PT (2011) Techno-economic analysis of autotrophic microalgae for fuel production. *Appl Energ* 88: 3524–3531.

16. Ahmad F, Khan AU, Yasar A (2013) The potential of *Chlorella vulgaris* for wastewater treatment and biodiesel production. *Pakistan J Bot* 45: 461–465.

17. Tebbani S, Filali R, Lopes F, et al. (2014) CO_2 biofixation by Microalgae: automation process, John Wiley & Sons.

18. Biller P, Ross A (2011) Potential yields and properties of oil from the hydrothermal liquefaction of microalgae with different biochemical content. *Bioresource Technol* 102: 215–225.

19. Wett B, Buchauer K, Fimml C (2007) In energy self-sufficiency as a feasible concept for wastewater treatment systems, IWA Leading Edge Technology Conference, Singapore, Asian Water, 21–24.

20. Borowitzka M (2005) Culturing microalgae in outdoor ponds, In: *algal culturing techniques*, Andersen, Ed. Academic Press, NY, USA, 205–217.

21. Jones S, Davis R, Zhu Y, et al. (2014) Process design and economics for the conversion of algal biomass to hydrocarbons: whole algae hydrothermal liquefaction and upgrading, Department of Energy Bioenergy Technologies Office, US.

22. Lundquist TJ, Woertz IC, Quinn N, et al. (2011) A realistic technology and engineering assessment of algae biofuel production. *Energ Biosci I*, 1–153.

23. Jonker J, Faaij A (2013) Techno-economic assessment of micro-algae as feedstock for renewable bio-energy production. *Appl Energ* 102: 461–475.

24. Sheehan J, Dunahay T, Benemann J, et al. (1998) A look back at the US department of energy's aquatic species program: biodiesel from algae, National Renewable Energy Laboratory Golden, CO, 328.

25. Ellis TG (2004) Chemistry of wastewater, Encyclopedia of Life Support System (EOLSS).

26. Chen Y, Liu J, Ju YH (1998) Flotation removal of algae from water. *Colloid Surface B* 12: 49–55.

27. Hansel PA (2014) Efficient flocculation of microalgae for biomass production using cationic starch. *Algal Res* 5: 133–139.

28. Jazrawi C, Biller P, Ross AB (2013) Pilot plant testing of continuous hydrothermal liquefaction of microalgae. *Algal Res* 2: 268–277.

29. Jena U, Das K, Kastner J (2011) Effect of operating conditions of thermochemical liquefaction on biocrude production from Spirulina platensis. *Bioresource Technol* 102: 6221–6229.

30. Berglin EJ, Enderlin CW, Schmidt AJ (2012) Review and assessment of commercial vendors/options for feeding and pumping biomass slurries for hydrothermal liquefaction, Pacific Northwest National Laboratory, Office of Scientific & Technical Information Technical Reports.

31. Garcia AL, Vos MP, Torri C, et al. (2013) Recycling nutrients in algae biorefinery. *Chem Sus Chem* 6: 1330–1333.

32. Faeth JL, Valdez PJ, Savage PE (2013) Fast hydrothermal liquefaction of *Nannochloropsis sp.* to produce biocrude. *Energ Fuel* 27: 1391–1398.

33. Holliday RL, King JW, List GR (1997) Hydrolysis of vegetable oils in sub-and supercritical water. *Ind Eng Chem Res* 36: 932–935.

34. Jena U, Vaidyanathan N, Chinnasamy S (2011) Evaluation of microalgae cultivation using recovered aqueous co-product from thermochemical liquefaction of algal biomass. *Bioresource Technol* 102: 3380–3387.

35. Elliott DC, Biller P, Ross AB (2014) Hydrothermal liquefaction of biomass: developments from batch to continuous process. *Bioresource Technol* 178: 147–156.

36. Elliott DC, Neuenschwander GG, Hart TR (2009) Catalytic hydrothermal gasification of lignin-rich biorefinery residues and algae, Pacific Northwest National Laboratory Pnnl.

37. Elliott DC (2008) Catalytic hydrothermal gasification of biomass. *Biofuel Bioprod Bioref* 2: 254–265.

38. Cruz FE, Oliveira JS (2008) Petroleum refinery hydrogen production unit: exergy and production cost evaluation. *Int J Thermodyn* 11: 187–193.

39. Kim S, Dale BE (2002) Allocation procedure in ethanol production system from corn grain I system expansion. *Int J Life Cy Assess* 7: 237–243.

40. Juneja A, Kumar D, Murthy GS (2013) Economic feasibility and environmental life cycle assessment of ethanol production from lignocellulosic feedstock in Pacific Northwest US. *J Renew Sust Energ* 5: 023142.

41. Frank ED, Elgowainy A, Han J, et al. (2013) Life cycle comparison of hydrothermal liquefaction and lipid extraction pathways to renewable diesel from algae. *Mitig Adapt Strat Gl* 18: 137–158.

42. King J, Holliday R, List G (1999) Hydrolysis of soybean oil. in a subcritical water flow reactor. *Green Chem* 1: 261–264.

43. Juneja A, Ceballos RM, Murthy GS (2013) Effects of environmental factors and nutrient availability on the biochemical composition of algae for biofuels production: a review. *Energies* 6: 4607–4638.

44. Sun A, Davis R, Starbuck M, et al. (2011) Comparative cost analysis of algal oil production for biofuels. *Energy* 36: 5169–5179.

45. Benemann JR, Oswald WJ (1996) Systems and economic analysis of microalgae ponds for conversion of CO_2 to biomass, Final report, California Univ., Berkeley, CA, Dept. of Civil Engineering.

46. Xiang X (2013) Techno-economic analysis of algal lipid fuels, Dissertation, Oregon State University.

47. Richardson JW, Johnson MD, Outlaw JL (2012) Economic comparison of open pond raceways to photo bio-reactors for profitable production of algae for transportation fuels in the Southwest. *Algal Res* 1: 93–100.

48. Palou RI, Wang MQ (2010) Updated estimation of energy efficiencies of US petroleum refineries, Argonne National Laboratory, US.

49. Frank E, Han J, Palou RI, et al. (2011) Life-cycle analysis of algal lipid fuels with the greet model, Center for Transportation Research, Energy Systems Division, Argonne National Laboratory, Oak Ridge.

50. Jorquera O, Kiperstok A, Sales EA (2010) Comparative energy life-cycle analyses of microalgal biomass production in open ponds and photobioreactors. *Bioresource Technol* 101: 1406–1413.

51. Weissman JC, Goebel R (1987) Design and analysis of microalgal open pond systems for the purpose of producing fuels: a subcontract report, Solar Energy Research Inst., Golden, CO., USA.

52. Xu X, Song C, Wincek R, et al. (2003) Separation of CO_2 from power plant flue gas using a novel CO_2 "molecular basket" adsorbent. *Fuel Chem Div Prepr* 48: 162–163.

3

Design of an off-grid hybrid PV/wind power system for remote mobile base station

Mulualem T. Yeshalem and Baseem Khan *

School of Electrical & Computer Engineering, Hawassa University Institute of Technology, Hawassa, 05, Ethiopia

* **Correspondence:** Email: baseem.khan04@gmail.com.

Abstract: There is a clear challenge to provide reliable cellular mobile service at remote locations where a reliable power supply is not available. So, the existing Mobile towers or Base Transceiver Station (BTSs) uses a conventional diesel generator with backup battery banks. This paper presents the solution to utilizing a hybrid of photovoltaic (PV) solar and wind power system with a backup battery bank to provide feasibility and reliable electric power for a specific remote mobile base station located at west arise, Oromia. All the necessary modeling, simulation, and techno-economic evaluation are carried out using Hybrid Optimization Model for Electric Renewable (HOMER) software. The best optimal system configurations namely PV/Battery and PV/Wind/Battery hybrid systems are compared with the conventional stand-alone diesel generator (DG) system. Findings indicated that PV array and battery is the most economically viable option with the total net present cost (NPC) of $57,508 and per unit cost of electricity (COE) of $0.355. Simulation results show that the hybrid energy systems can minimize the power generation cost significantly and can decrease CO_2 emissions as compared to the traditional diesel generator only. The sensitivity analysis is also carried out to analysis the effects of probable variation in solar radiation, wind speed, diesel price and average annual energy usage of the system load in the optimal system configurations.

Keywords: Base Transceiver Station (BTSs); photovoltaic (PV); wind power system

1. Introduction

Stand-Alone Systems are designed and optimized to meet the power demand of remote places are off-grid power systems. An off-grid system does not have a connection to the main grid electricity and vary widely in size and application [15].

Hybrid power systems are designed for the generation of electrical power using number of power generation devices such as wind turbine, PV, micro hydro and/or other conventional generators. In addition, it includes power electronics and electricity storage bank. Some of the advantages of using RESs are gain an immediate access to reliable electricity at any time; reduce the dependency from oil price fluctuations and the transportation costs of fuels; increase economic productivity and fight climate change[16–19].

Telecommunication network have changed the way people live, work and play. Since many people around the world are connected by mobile networks, the challenge to provide reliable and cost effective power solutions to these expanding and remote networks is indispensable for telecom operators. The mobile Industry in Africa faces many challenges to operate the mobile networks in a cost effective manner. Africa has one of the lowest electrification rates in the world with only 43% of the population having access to grid electricity. Wherever there is access to grid electricity, the supply of electricity is highly unreliable with frequent and long outage [1].

Africa is a land of renewable energy source's opportunity. However, currently only less than 2% of the renewable resources (excluding hydro) are exploited for electricity generation across Africa. This presents a huge untapped potential for large scale renewable power projects as well as small scale mini-grids and off-grid power systems [20].

Africa has one of the lowest electrification rates in the world with only 43% of the population having access to grid electricity. Wherever there is access to grid electricity, the supply of electricity is highly unreliable with frequent and long outage [20,21].

Despite the fact that around 80% of the population of Ethiopia lives in rural areas, electricity supply from the grid is almost entirely concentrated in urban areas. And limited grid infrastructure and inadequate power generation capacities has greatly affected the availability and quality of electricity supply. Among other things, dispersed demand and very low consumption level of electricity among rural consumers, limited grid electricity penetration to rural population.

The latest national energy balance indicates that Ethiopia consumed 1.3EJ of energy in 2010. This was derived from biomass fuels (92%), hydrocarbons (7%), and electricity (1%). The main consumers of energy were the residential and service sector (93%) and transport (5%) with the remainder going for industrial and other applications [6].

In Ethiopia, the mobile phone ownership/ mobile network coverage has reached 24.7% of households in 2011—65.2% of households in urban areas and 12.8% of households in rural areas. In contrast rural household electricity connection was only 4.8% in 2011. Mobile phone owning households in rural areas had already reached 1.75 million in 2011. Rural mobile ownership can be expected to reach 45% (this is the planned mobile network coverage for 2015) or 6.8 million households by 2015. Since the electrification rate is going much slowly than mobile phone network coverage more than two third of rural mobile phone owners (or more than 4 million households in 2015) would not be connected to electricity [6]. The electric power infrastructure is playing a negative role in the growth of mobile telecommunications in terms of network coverage and great impact on the operation cost of running the system due to non-availability of grid power supply. Due to this there is very limited or no coverage of mobile network for the rural population.

In the conventional diesel generators with backup battery were used for powering these mobile tower sites (BTS). These off-grid systems, usually located in areas with difficult accessibilities require regular maintenance and are characterized by their high fuel consumption and high transportation and operational cost [10,11]. Also, due to the increasing demand of clean energy technology to reduce the greenhouse gas emission pressure telecom companies for alternative solution for powering these sites is needed.

Therefore, in order to meet the continuous typical load demand of a mobile base station during varying atmospheric conditions, different energy sources need to be integrated for extended usage of alternative energy. This will create a large demand for off-grid power supply in rural areas which renewable energy is best suited to realize ethio-telecom tower site with renewable energy technologies. Solar and wind are available freely and thus appears to be a promising technology to provide reliable power supply in the remote areas and telecom industry of Ethiopia. The project aim to design an off-grid hybrid renewable energy system for Base Transceiver Station (BTS), so that can generate and provide cost effective electric power to meet the BTS electric load requirement.

1.1. Literature review

Different research is carried in the field of renewable energy; applications of stand-alone power system and hybrid renewable energy systems have been conducted for the maximum usage of the resource potential.

A case study [7] in the Somalia region of Ethiopia, The remote rural village called Werder district (6050'N 45030' E) have an average wind speed of 5m/s at 10m elevation and an average daily solar radiation of 7.5kwh/m2/day. Extension of national grid is not economically feasible and the electrical load density in the village is low. The techno economic analysis of this option has been done using HOMER software and hybrid PV/wind/diesel generator system became economically feasible for the proposed site based on some important parameters such as high renewable penetration, less annual diesel consumption, less carbon dioxide emission and less cost of energy.

The study [10] presents the result of techno-economic analysis of hybrid system comprising of solar and wind energy for powering a specific remote mobile base transceiver (BTS) in Kaduna state, Nigeria. But the optimal system configurations obtained through simulation in HOMER. Two best optimal system configurations namely PV-Diesel-Battery and PV-wind-diesel-battery system are compared with the conventional stand-alone diesel generator (DG) system. Finding indicated that PV array (10KW)-DG (5.5KW)-Battery (64 units Trojan L16P) is the most economically viable option with the total net present cost of $69,811 and per unit cost of electricity of $0.409. The simulations indicate that a hybrid system option, compared to a diesel only system, is feasible for each of the three villages.

In the paper [14] they are proposed a hybrid system cost analysis which has wind generation, solar system, and storage battery system and diesel generator using efficient optimization tool HOMER for obtain the optimal cost of the hybrid system. Determine the optimal combination of solar, wind and diesel based hybrid system to fulfill the load requirement and minimize the cost of telecommunication site in BSNL Bhopal, India.

In this paper [11] presents a solution utilizing a hybrid of solar and wind power systems with a portable generator to provide reliable power for a mobile base station located behind the Himalayas of south Asia. The design is based on the local mobile subscribers with 350 capacities with 51mE per subscriber traffic and the peak load capacity of 750 W. The meteorological data including solar sunshine hours and hourly wind speed are taken for a site in Mustang district at 3444 m altitude. The power consumption patter of a mobile base station depends up on the traffic pattern of the mobile users. The cost of the hybrid system is also estimated as $81,512.04 Canadian dollars. The proposed system ensures the reliability of power supply to run the 24/7 cellular mobile services at an extremely remote site of Nepal.

Thus, based on the literature reviews, HOMER modeling software is taken for the purposes of this study to carry the feasibility analysis. when compared with other software's HOMER creates a list of feasible system configurations sorted according to cost effectiveness and presents the optimal configuration based on the lowest net percent cost (NPC) the supply in order to design system [23]. This paper also used the same software to design and optimize the off-grid hybrid power system to be provided electricity requirement of the remote telecom site in Ethiopia.

To improve the communication infrastructure of the rural community the ethio-telecom must use Hybrid RES to provide electrical power depending on the geographical area and the resource availability of the area. Here in this site, it uses Hybrid Battery and Generator, but power supply is not continuous. However this paper differs from the related studies in terms of application, load demand, climatic data, and location of the study area.

2.   Background of the Study Area

2.1. Energy Potential of the Study Area

The mobile telecom base station considered for this hybrid system project is located in Ethiopia in Oromia Region of West Arsi, with Geographical coordinates of latitude 7.20592 and longitude 38.60801. Here, the mobile telephony base station is taken from ethio telecom site; the global system for mobile (GSM) and code division multiple access (CDMA) network system base station is considered. Since Ethio telecom is the only operating companies in Ethiopia.

Ethiopia is located near the equator, there is a significant potential of solar resource. Solar resource raw data input to the HOMER software is the average global horizontal radiation obtained for the site from Atmospheric Science Data Center (ASDC)-NASA surface meteorology and solar energy database [12].

Table 1. Average monthly global solar radiation of West Arsi, Oromia region of Ethiopia.

Month	Jan	Feb	Mar	Apr	May	Jun	Jul	Aug	Sep	Oct	Nov	Dec	Ave.
Solar radiation (kWh/m2/day)	6.02	6.41	6.35	6.04	5.95	5.42	4.83	5.01	5.64	6.04	6.25	6.10	5.83
Clearness index	0.652	0.652	0.614	0.577	0.581	0.541	0.479	0.486	0.547	0.609	0.669	0.677	0.588

From Table 1 it can be seen that the monthly average solar energy resource is well distributed in the site with average monthly solar radiation of 5.834 kWh/m^2/day and average daily sunshine hours of 11 hr. The maximum solar radiation is for the month of February having daily radiation of 6.41 kWh/m^2/day, February was the sunniest month of the year, whereas the minimum value is occurred during the peak rainy season months (July and August) in the country, particularly in the month of July with radiation of 4.83 kWh/m^2/day. It also depicts the clearness index of the site obtained after HOMER simulation. The clearness index tells about the clearness of the sky from the latitude and longitude of the site considered. Here in this study the clearness value varies from 0.479 in July to 0.677 in December.

Note that HOMER assumes the output of the PV array is linearly related to the solar radiation incident on the PV array, and also ignores the effect of ambient temperature on the performance of the PV array. The nominal operating cell temperature is the surface temperature that the PV array would reach if it were exposed to 0.8 kW/m^2 of solar radiation, an ambient temperature of 20 °C, and a wind speed of 1 m/s.

The clearness index in HOMER tells about the clearness of the sky of the site, meaning the transmission of the radiation directly from sky to earth's surface. The clearness index value is dimensionless and varies from 0 to 1 representing the cloudiest and sunniest months respectively and there is an average daily sunshine hours of 11 hr as shown in Figure 1.

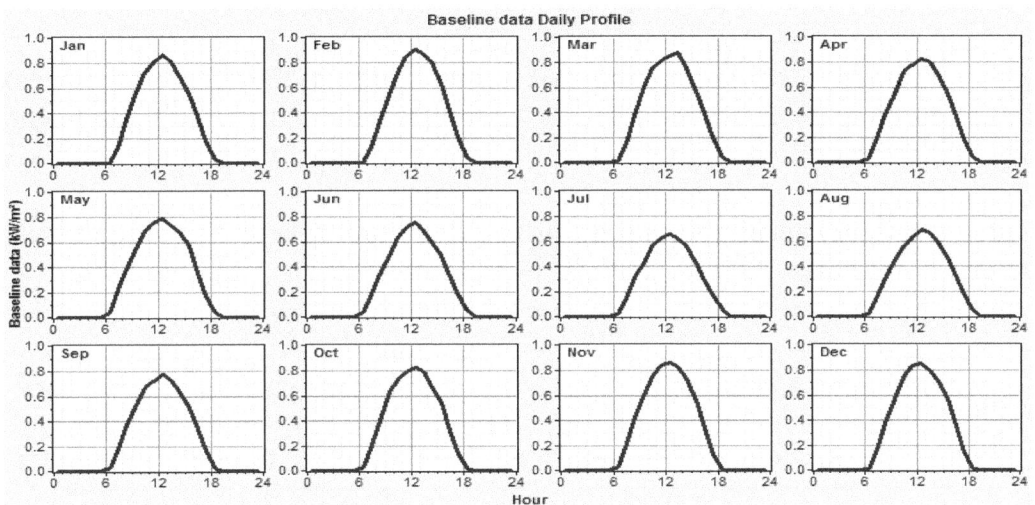

Figure 1. Diurnal Variation of Global Horizontal Solar Radiation Source.

Normally from middle of June till end of August is summer season in region which means it is rainy time but the solar radiation is enough and a substantial amount of electricity could be generated, thus the considered site in this study has shown excellent solar energy sources to be exploited to generate electricity. From HOMER analysis, it can be seen that the considered site has enormous potentials for PV applications.

There are two basic windy areas in Ethiopia located alongside the main East African Rift Valley, the North Eastern highlands of the country near Tigray regional state, the southern part of Ethiopia near the Kenyan border, the central Ethiopia and eastern lowlands part of Ethiopia, however, a significant amount of it had not been harnessed yet throughout the county [7,8].
Monthly mean wind speed resources of the site are obtained from NASA surface meteorology and solar energy database. NASA has estimated the annual average wind speed of the location to be 2.97 m/s at an anemometer height of 10 m [9,12] as shown in Table 2.

Generally it shows the wind speed data of the site is not that much satisfactory for power generation since it has a very small cut-in wind speeds. So, wind speed extracted from NASA is simply taken to assess wind energy potential of the selected site (resource assessment). This data can be extrapolated to the designated wind turbine height of 30 m. Tables 2 summarize the monthly wind speed variation of the site at 30 m heights where the average wind speed found to be 3.687 nm/s. The lowest wind speed profile is indicated in the months of July, August and September; Whereas, November, December and January are the windiest months.

As wind turbine height increase length from ground the wind speed also increase and hence, power production also rises. The wind speed at any height above ground level can be interpolated either in exponential function or logarithmic function forms [24]. In this paper the logarithmic function used that the logarithmic profile (or log law) assumes that the wind speed is proportional to the logarithm of the height above ground. The following equation therefore gives the ratio of the wind speed at hub height to the wind speed at anemometer height:

$$\frac{V_2}{V_1} = \frac{In(\frac{h_2}{z_0})}{In(\frac{h_1}{z_0})}$$

Where: Zo: Surface roughness length factor [m]; Surface roughness length describes the roughness of the surroundings terrain in this case few trees, having a value of 0.01. h1: Reference height above ground level [m], h2: Hub height [m].

Table 2. Average monthly wind speed at 10m and 30m Height of West Arsi, Oromia region of Ethiopia.

Month	Jan	Feb	Mar	Apr	May	Jun	Jul	Aug	Sep	Oct	Nov	Dec	Ave.
Wind Speed (m/s) at 10m	3.55	3.21	3.03	3.10	2.94	2.90	2.52	2.34	2.41	2.89	3.32	3.53	2.97
Wind Speed (m/s) at 30m	4.397	3.976	3.753	3.636	3.840	3.592	3.121	2.898	2.985	3.579	4.112	4.372	3.687

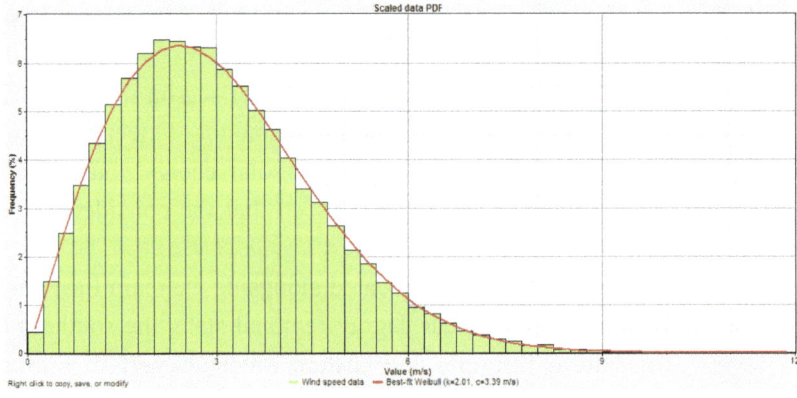

Figure 2. Probability Distribution Function of Wind Speed data of West Arsi, Oromia region of Ethiopia.

From this Figure 2 it can be seen that the most probable wind speed range 2.0–3.5 m/sec occurs approximately 35%, and the wind speed above 3.5 m/s is occurred around 45% of the time. Thus this shows that some of the wind energy could be exploitable.

The characteristic curve of Bergey wind turbine, which exhibits wind turbine power output variation with wind speed, the cut out wind speed is around 8 m/sec and cut in speed is more than 4 m/sec at hub height as shown in Figure 3.

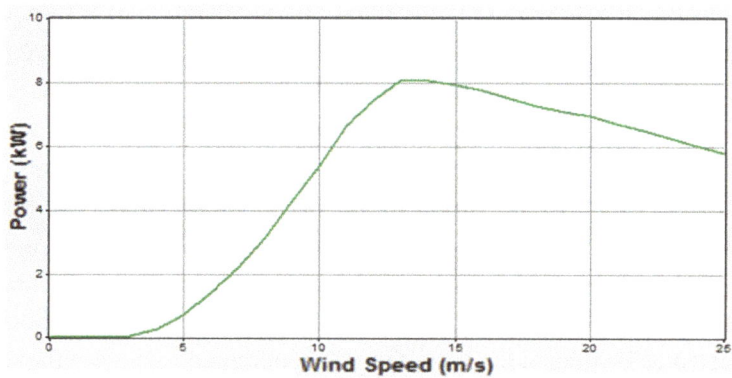

Figure 3. Power Curve of Bergey wind turbine.

2.2. Analysis of The BTS Electricity Load

The main electrical and electronics equipment of this mobile network site are Radio Base Station (RBS), Power Base Controller (PBC) including Rectifier, Battery Base Station (BBS) and Diesel Generator (DG) with Fuel Tank [2,3]. Typically, a conventional BTS site load consists of BTS equipment load as well as air conditioner and lighting loads. So A second-generation GSM & WCDMA System mobile base station consisting of a single RBS 6101 with three RF Antenna at angle of 120 at 29 m height; with one MW Antenna at 36.6 m high, with Lighting Rod and Aviation Light at 40 m high; one PBC 05 with 3 Rectifier of 2 KW Power rating; BBS 6101 with 8 Battery per two battery Strings and DG with Rating of 10 KVA and Fuel tank with 1000 L Capacity was considered in this study [4,5] This equipment's are outdoor Material. Diesel generator is scheduled to operate accordingly for the whole day throughout the year. Weather it is high or low traffic demand, DG is scheduled to run in optimal conditions for the rest of hours.

Most of the radio transceiver appliances employed in mobile network site use DC power to operate. There AC power supply from the DG to the AC input terminal of rectifiers to convert in DC power and transfer to the DC loads. The DC load components are connected to −48 V DC power supply.

More than 60% of the power is consumed by the radio equipment and amplifiers, 11% is consumed by the DC power system and 25% by the cooling equipment, an air conditioning unit [5]. An individual mobile phone tower, the BTS can account for approximately 4 kW–6 kW of total energy consumption. Comprising BTSs (the primary radio equipment), an air conditioner (if required) and antennas and lighting, are the largest energy consumers at a tower site.

2.3. Load Pattern for Mobile Base Station

The Load estimation is assumed based on the electric and electronic appliances used in the mobile network site and usage of the electrical energy; because the power consumption pattern of a BTS depends up on the number of subscribers, the area to be covered, the topography of this area, and the technology used.

Although, measured hourly load data for the considered site is difficult to find the exact hourly load usage data. However, it is important to note that, the stated rating of BTS equipment will not be loaded to full capacity at all times due to the variability in operating hour of the equipment.

Normally, the mobile network load requirement to be the same for entire days of the year and there is low variation in the energy load profile requirements of a BTS across the day. The load can be classified into two categories. The load is low throughout the night till morning, whereas in the busy category occurred usually occurs during in the day time particularly the business hours of the day around lunch time took the maximum power demand. Therefore, BTS is operated at busy load and low load condition under maximum and minimum traffic. It should also be noted, that the air conditional is considered to only operate during the business hour (sunshine hour) of the day when high power is expected to be demanded from the BTS equipment, whereas the miscellaneous loads (Aviation lighting & florescent lamp) is operated only in the night hours.

The hourly load profile of the site for a typical day (1 January) is shown in Figure 4 from where scaled average energy consumption per day, scaled daily peak demand, and daily average demand are found to be approximately 41.4 kWh, 3.01 kW, and 1.72 kW, respectively. The mismatch in the peak value and value shown in Figure 4 is due to the scaling process and the random variability introduced in HOMER to make the load pattern unique. Based on this variation, a day-to-day random variability of 5% and hour-to-hour random variability of 10% was specified in HOMER so as not to underestimate the peak load the proposed system can serve.

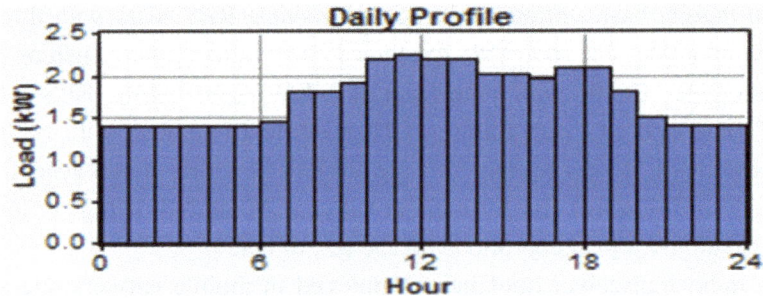

Figure 4. Daily load profile of the site.

To synthesize data in HOMER, you must enter at least one load profile, which is a set of 24 hourly values of electric load, one for each hour of the day. You can enter different load profiles for different months and for weekdays and weekends too. But if you only enter one load profile it will be used throughout the year. HOMER adds randomness according to the values you enter for daily noise (5% for day to day Variation) and hourly noise (10% for hour to hour Variation) and calculates the average 24-hour load profile for the whole year.

3. Design of the Proposed System

The proposed energy system should meet the load demand of the site. The main renewable sources of energy considered in this project are solar and wind. The diesel generator produces AC type voltage as backup, whereas the PV panels and wind turbine output is DC type. The converter is added to maintain the flow of energy between the AC and DC components, whereas the battery is employed as energy storage systems in order to ensure uninterrupted power and to maintain the desired power quality at the load point because of unpredictable variation in the climatic condition affect nature of the renewable sources. Hybrid model of these three energy sources in parallel with battery storage make the hybrid power system more reliable, efficient and provides a smooth and uninterrupted power supply. Figure 5 presents the schematic representation of HOMER simulation model considered.

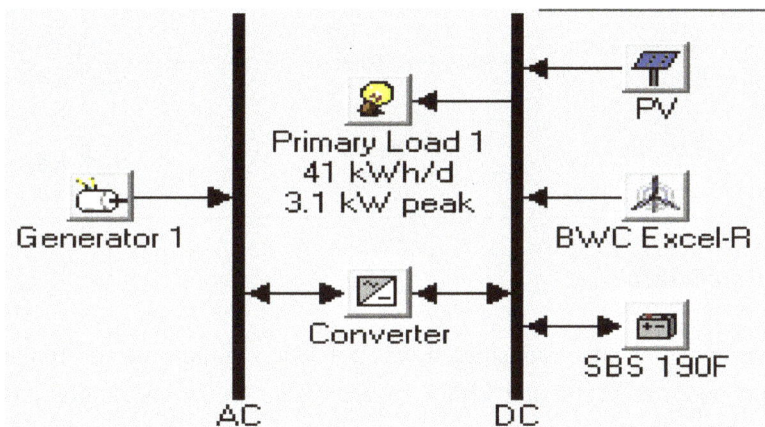

Figure 5. The proposed hybrid system produced by HOMER.

In the proposed system, dispatch strategies are considered which means HOMER will simulate each optimal system to determine to charge/discharge of batteries and control the operation of the diesel generator. In such a system, the battery bank absorbs energy when the renewable energy output exceeds the load and discharges energy when the load exceeds the renewable output. And one of the Constraint inputs assumed in HOMER are the system select the large amount energy from renewable fraction compare with diesel generator based on the cost.

4. Materials Cost and Size Specification

The HOMER software is used to determine the best optimal sizing and feasibility study of the system. The initial choice of the components size is based on the site load profile. Some of the input values into the software are expressed in size and in quantity. Wind turbines, batteries are the power system components which vary in quantity, and solar PV, diesel generator and converter are other components that vary in size. The main purpose of the work is searching the optimum power system configuration that would meet the load demand with minimum NPC and COE. The brief description of the main components of the proposed hybrid system is summarized in Table 3; the basic criterion related to the selection of the power system components are the overall cost of individual materials.

So we use the current cost of all the necessary components from different companies around the end of 2015.

Table 3. Summary of cost and size of components.

No	Component	Size(KW) Qty. (No.)	Capital Cost($)	Replacement Cost ($)	O&M Cost ($/Year)	Consider size(kW)	lifetime
1	Solar PV	1kW	$2500	$2000	25	0, 1, 2, 3, 4, 5, 6, 7, 8, 9,10	20 year
2	Wind Turbine	7.5kW	$16870	$11809	$337	0, 1, 2, 3, 4, 5	20 year
3	Diesel Generator	8kW and 10kVA	$5700	$570/ kW	$0.35/h	0, 3, 5, 8 10	15,000 hr.
4	Battery 4batteries in one string	6 cells, 12V, 190Ah, 2.3 kWh	$300	$300	$10/year	0, 8, 12, 16, 20, 24, and 28	1000 kWh
5	Power Converter	1kW	$650,	$650	$10/year	0, 2, 4, 6, and 8 kW	10 years

4.1. Solar PV Size and Cost

In this project the solar panel considered was a 1 kW, capital cost and replacement cost for 1 kW of PV array are taken as $2500 and $2000, respectively. As the installation cost is taken as 60% of the PV price and the operation and maintenance cost would be 1% per year [10]. In this case, 10 different types of PV arrays are considered to get the optimal size (0, 1, 2, 3, 4, 5, 6, 7, 8, 9, and 10 kW). The lifetime of PV arrays was taken as 20 years and no tracking system was considered. The following default parameters were considered for modeling of the power system like; the de-rating factor was taken as 90%, ground reflectance was also considered as 20%, slope 7.2 and azimuth 0 (south orientation). The PV derating factor is a scaling factor that HOMER applies to the PV array power output to account for reduced output in real-world operating conditions for such factors as ambient temperature, soiling of the panels, wiring losses, shading, snow cover, aging, and so on.

4.2. Wind Turbine Size and Cost

The wind turbine is manufactured by Bergey Wind Power having the model BWC Excel-R with a rated capacity of 7.5 kW and provides DC is used in this project. The initial cost of one unit in the current market price is considered as $16870. Replacement and annual operational maintenance costs were assumed as $11809 and $337year, respectively. O&M cost of wind turbine was proposed about 2% of its initial capital cost as given in [13]. Replacement cost of the wind turbine considered in this case is about 70% of capital cost after 20 year service life. In order to find an optimal size, five different wind turbine options were analyzed: 0, 1, 2, 3, 4, and 5 turbines. The operational lifetime of a turbine is considered as 20 years and Hub height is 30m. As we know the advantage of Bergey wind turbines is easy to install particularly in remote telecom sites and the tower height can be adjusted to the required hub height because the tower is metal most of the time so good to construct and mount the mobile tower mechanically.

4.3. Diesel Generator Size and Cost

The cost of diesel generators available in the market varies but the DG used in the in particular Ethio-telecom BTS site is Deutz air cooled DG with rate output power of 8 kW and 10 kVA with the initial cost of $5700 and the fuel consumption at full load is less than 3 litters. The initial capital cost of the DG is assumed $570/kW. Replacement and operational costs were assumed $570/kW and $0.35/h, respectively. The operating lifetime was also considered 15,000 h. currently, per liter price of diesel in Ethiopia is around $0.8. In this study, 0, 3, 5, 8 and 10 kW sizes of DGs were also considered for simulation. The minimum load ratio was set at 10%.

4.4. Battery Size and Cost

The lead acid battery of EnerSys SBS model battery with 6 cells, 12 V, 190 Ah, and 2.3 kWh is selected in the proposed system, since this the type of battery used in the selected Ethio-telecom BTS site with 4 batteries in one string of 48V and total of 3 stings. The initial cost of one unit is considered as $300. Replacement and operational maintenance costs were assumed as $300 and $10/year, respectively. In order to find an optimal configuration, the battery bank was assumed to consist of any number of batteries (0, 8, 12, 16, 20, 24, and 28). Each battery string contains four batteries and the lifetime throughput of each battery is estimated to be 1000 kWh.

4.5. Power Converter Size and Cost

The initial capital cost and replacement cost and operation-maintenance cost of the convertor that is considering in system is 1 kW are $650, $650 and $10/year respectively. The efficiency of the converter is assumed to be 90% and the lifetime is considered as 10 years. Different sizes of converters (0, 2, 4, 6, and 8 kW) are considered during analysis using HOMER.

5.   Results and Discussion

5.1. Optimization and Simulation Results

After introducing all of the input variables, the HOMER software is run repeatedly to get the feasible system configuration; component sizes that meet the load requirement and the inputs constraints at the lowest NPC and then presents the results of the simulation in terms of optimal systems and sensitivity analysis. The Optimization result displays ten different configurations according to the lowest NPC for further analysis and to increase the chance of finding most optimized system. Hybrid system with less NPC, less COE, higher renewable fraction, less capacity shortage, and minimum fuel consumption would be suggested as optimum system.

HOMER simulates every system in all set of component combination (search space) and ranks all the feasible systems according to increasing net present cost. In the case of this paper it has 81 sensitivity performing 23760 simulations.

	PV (kW)	XLR	DG (kW)	SBS 190F	Conv. (kW)	Disp. Strgy	Initial Capital	Operating Cost ($/yr)	Total NPC	COE ($/kWh)	Ren. Frac.	Capacity Shortage	Diesel (L)	DG (hrs)
	9			20		CC	$ 28,500	2,489	$ 57,508	0.355	1.00	0.09		
	7	1		20		CC	$ 40,370	2,461	$ 69,050	0.426	1.00	0.09		
	10		3	28	1	LF	$ 35,760	3,421	$ 75,631	0.434	0.97	0.01	245	491
	8	1	3	28	1	LF	$ 47,630	3,397	$ 87,217	0.499	0.97	0.01	245	494
	8		3		2	LF	$ 23,010	10,189	$ 141,751	0.807	0.64	0.01	3,605	5,933
		1	3	8	4	CC	$ 23,580	11,064	$ 152,518	0.866	0.22	0.00	4,510	4,559
	7	1	3		2	LF	$ 37,380	9,951	$ 153,343	0.871	0.70	0.00	3,183	5,664
			3	12	4	CC	$ 7,910	13,075	$ 160,278	0.910	0.00	0.00	5,826	5,884
			3		2	CC	$ 3,010	15,181	$ 179,922	1.053	0.00	0.07	6,175	8,760
		1	3		2	LF	$ 19,880	14,505	$ 188,919	1.085	0.23	0.03	5,267	8,411

Figure 6. Optimization results of the system.

The two best optimal hybrid system configurations PV/Battery and PV/WT/Battery which ranks first and second by the net present cost (NPC) with levelized cost of energy (LCOE), will be compared with the conventional DG/Battery/Converter system rank in eight. The performance analysis of the two selected hybrid systems configuration and standalone DG system are discussed in the proceeding sections. The parameters used for the comparison and analysis of hybrid power systems are NPC, COE, Renewable fraction, fuel consumption, capacity shortage, and excess electricity generation as shown in table 4.

It can be seen from this table 4 below that the site can adequately rely on the RE source to power the remote telecom BTS towers equipment due to high presence of solar radiation and wind speed at the site.

Table 4. Performance parameters of three different system configurations.

Models / Parameters	PV/BB System (First Choice)	PV/WT/BB Systems (Second Choice)	DG/BB/Converter System (Existed system)
PV size (kW)	9	7	0
Wind Turbine (quantity)	0	1	0
Generator Size (kW)	0	0	3
Battery Bank (Quantity)	20	20	12
Converter Size (kW)	0	0	4
Dispatch Strategy	CC	CC	CC
Initial Capital cost ($)	28,500	40,370	7910
M&O Cost ($/year)	2,489	2,461	13,075
Total cost of Fuel ($/year)	0	0	4,660
Diesel Fuel Used (L)	0	0	5,826
DG Running hours (hrs.)	0	0	5,884
Renewable fraction (%)	100	100	0
Capacity Shortage (%)	9	9	0
Excess Electricity (KWh/year)	2,405	2,512	0
Unmet load (kWh/year)	1,207	1,189	0
Total electricity production (kWh/year)	17,436	17,357	17,652

5.1.1. PV/ Battery configuration

The best optimal combination comprised of 9 kW PV array and 20 units EnerSys power safe SBS 190F battery at 5.83 kWh/m^2 per day average global solar radiation, 3.687 m/sec annual average wind speed and 0.8/L diesel price. It has a renewable fraction of 100% solar energy without inclusion of wind. The cost summary of this system has a total net present cost of $57,508, levelized cost of energy of $0.355/kWh and operating cost of $2,489 per year. The total NPC for the stand-alone diesel system is $160,278 which is 3 time higher than PV/battery configuration. The simulation result recommends that even a Solar-Battery System is enough to fulfill the demand for basic telecom load profile with excess of electricity 2,405 kWh/year. It can be seen that the site can adequately rely on solar energy to power the BTS loads due to high presence of solar radiation at the site as shown in Figure 7.

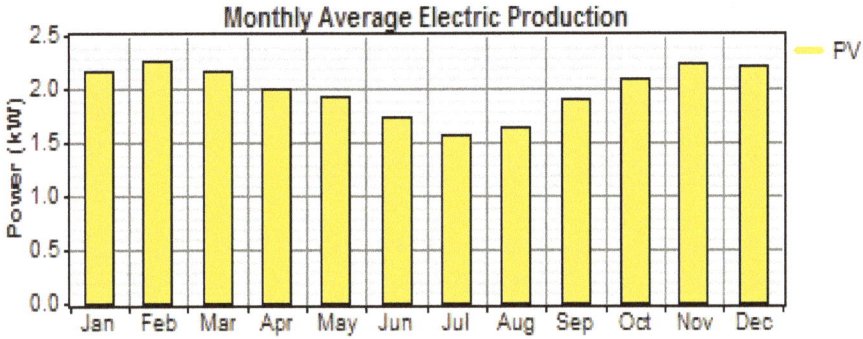

Figure 7. Monthly average electricity production of PV/Battery hybrid system.

5.1.2. PV/Wind/Battery configuration

The next optimal combination consists of 7 kW PV array, one 7.5 kW BWCXL-R wind turbine and 20 units EnerSys power safe SBS 190F battery and no converter because all electricity produced are DC. The result is based upon the system with 41.4 kWh/day telecom load at 5.83 kWh/m^2 of solar radiation, 3.687m/s of wind speed and $0.8/L diesel price.

The COE of this hybrid system is $0.426/kWh and the total NPC and Operating cost of the system were found to be $69,050 and $2,461 respectively. On the other hand, NPC of the stand-alone diesel system is more than 2 times higher than this configuration. The total average electric production of the hybrid system is found to be 17,357 kWh per year, of which PV array contribute 78% and wind turbine provides 22% of the total annual production.

But the consumption as per the dc telecom load is found to be 20,344 kWh per year resulting excess electricity of 3969 kWh per year with unmet electric load of 1556 kWh per year and capacity shortage of 1825 kWh per year.

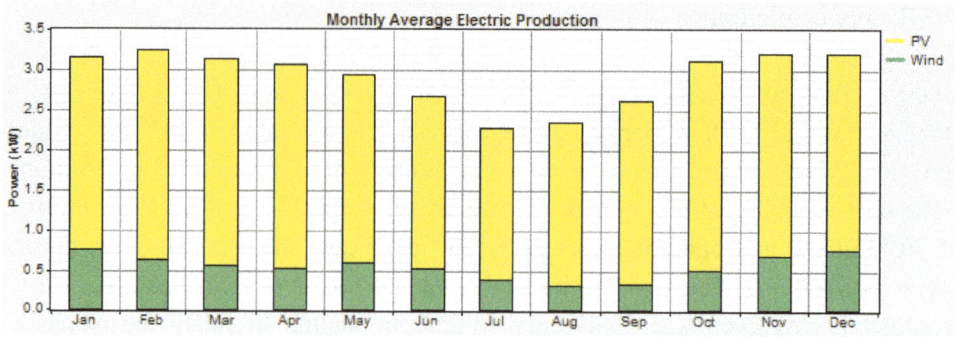

Figure 8. Monthly average electricity production of PV/Wind/Battery hybrid system.

As Figure 8 shows the PV system will operate more compared to the wind turbine, which means the PV array serves the base load. The electricity production from the wind turbine is limited due to the lower wind speed at the site, although the renewable energy fulfills the telecom load demand 100%.

5.1.3. DG/Battery/Converter configuration

The conventional standalone generator system consists of 3KW DG, 12 unit battery, and 4KW converter. This project reveals that DG-Battery System is very costly with cost of energy (COE) $0.910/kWh and takes large total NPC of $160,278 due to the huge operation cost of $13,075 per year when compared with other hybrid system. So, the DG NPC cost takes more than 85% of the total NPC cost and the NPC cost of the battery and the converter are around 10% and 5% of the total NPC cost respectively. The reason is the cost of Fuel and O&M are very large. On the other hand, the standalone diesel system consumes a total of 5,826 L/year and run for 5,884hr/year. The total electrical energy production from the DG system is found to be 17,652 kWh/year and DC primary load consumption of 15111 kWh per year; resulting no excess electricity production, no capacity shortage and zero unmet electric loads. CO_2 emission is found to be 15,341 kg per year followed by 37.9 kg/year of CO emission per telecom tower.

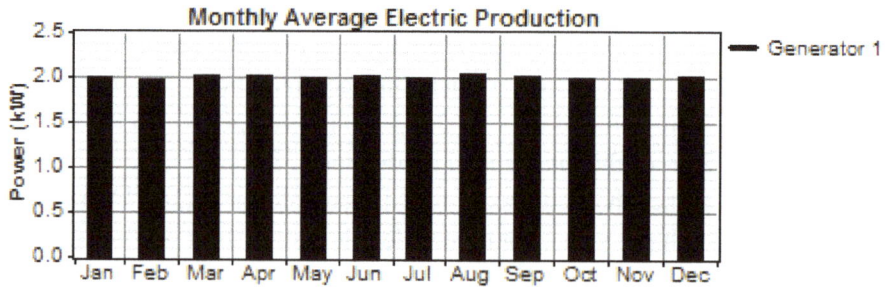

Figure 9. Monthly average electricity production of DG/Battery system.

5.2. Sensitivity Result

Sensitivity analysis helps in exploring the effect of the changes (uncertainty dynamic change like, increasing or decreasing demands, renewable and non-renewable resources fluctuations.) in the available resource and economic condition. This analysis shows the power system sensitive for each change of the input variables. Four sensitivity parameters were considered in this study, namely: solar radiation (3, 5.83 & 8 kWh/m2/day), wind speed (2, 3.687 & 6 m/sec), annual average energy (20, 41.4 & 60 kWh/day) and diesel price (0.4, 0.8 & 1.5$/L).

There are six different sensitivity results scenarios for the Optimal System Type (OST) they are: Variation in wind speed and solar radiation, Variation in wind speed and diesel price, Variation in solar radiation and diesel price, Variation in solar radiation and Load average energy, Variation in diesel price and Load average energy, and Variation in load average energy and wind speed.

5.2.1. Varying average wind speed and average solar radiation value

In this sensitivity analysis, OSTs are investigated for three different values of load average energy of BTS and diesel prices as presented in above of Sensitivity analyses that means we have a total of nine combinations for Optimal system types graph for which the wind speed range 3.0 and 6 m/sec in the X-axis and solar radiation range 4.0 and 7.0kWh/m2/day as shown in Figure 10.

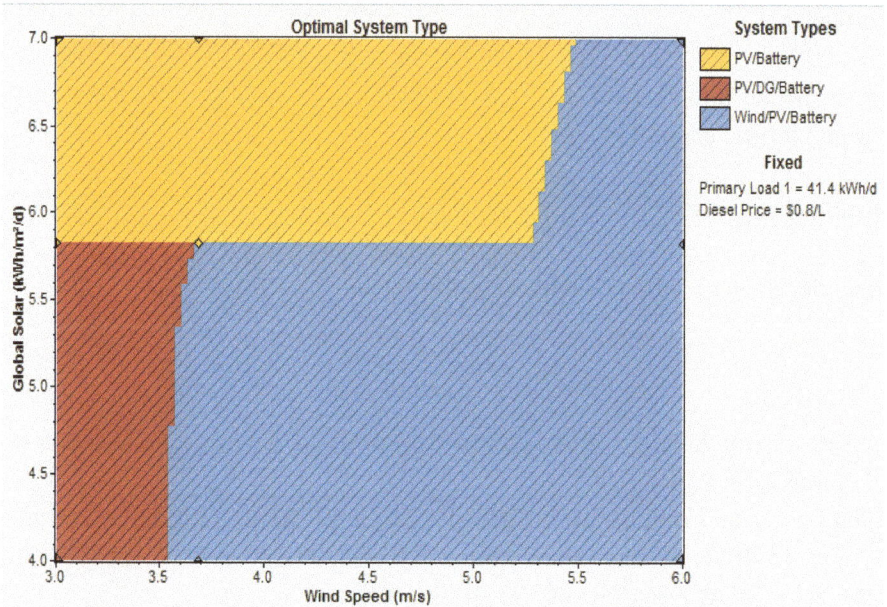

Figure 10. Optimal system types for varying solar radiation and wind speed values.

It is obvious from this Figure 10 that for the specified values of Primary load (41.4 kWh/day) and diesel price ($0.08/L) there is three best optimal configurations PV/Battery, Wind/PV/Battery and PV/DG/Battery at the considered site, therefore it is more economical.

In the remaining sensitivity values it can be observed that there are different configurations for the system. The sensitivity analysis showed that almost the same configuration was obtained except in some case there are quantity and size change of the components. The different setups resulted in this paper could be appropriate in areas that have the same climatic resources.

6. Conclusion

The paper indicates that the site is blessed with considerable annual average global solar radiation of 5.83 $kWh/m^2/day$ and average wind speed of 3.69 m/sec at 30 m height. Therefore; there is a potential site for installation of PV/Battery hybrid system and PV/Wind/Battery hybrid system. The simulation results from HOMER show that the most economically feasible configuration for the site; both configurations have a renewable penetration of 100%. However, the conventional standalone diesel configuration is not economically due to the high cost of fuel consumption and O&M, which might have significant effect on the operating cost of the mobile telecom operation and also environmental impact due to high CO_2 & CO emission from the system.

Therefore, utilizing renewable-energy based hybrid system for powering BTS sites will decrease the operating cost of the telecom company which has direct impact on the low-income rural telecoms subscriber, but also eliminate the diesel fuel usage, thereby bring low greenhouse gas emission to the environment. Finally, the information gained from the entire project can be applied in the design, execution, or development of HRES for any mobile tower applications in other locations in the country.

Acknowledgment

The authors would like to thank the Ethio-telecom and Hawassa University for providing the enabling environment and support to carry out this work.

Conflict of Interest

All authors declare no conflict of interest in this paper.

References

1. IFC, Green Power for Mobile (2014) Tower Power Africa, Energy Challenges and Opportunist for Mobile Industry in Africa.
2. Site Installation documentation, SID NSB ASP 222626 SR REV, ERICSSON, ethio-telecom.
3. Ericsson AB 2013/ RAN System Description and RBS6000 Survey/LZU1088684 R3Apdf.
4. Ericson AB 2015/ MOD RSM Low Level Design pdf.
5. Ericsson AB 2010 RBS 6102/6101 & BBS/BBU/GBF 6102/1 6101 Installation 2010-08-05 pdf.
6. International Solar Energy Institute, Germany/Ethiopia, shifting solar energies-Solar Energy Foundation (2012) Solar energy vision for Ethiopia, Opportunities for creating a photovoltaic industry in Ethiopia, ISEI.
7. Tesema S, Bekele G (2014) Resource Assessment and Optimization Study of Efficient Type of Hybrid power System for Electrification of Rural District in Ethiopia. *Int J Energ Power Eng* 3: 331-340.

8. Abraha AH, Kahsay MB, Kimambo CZM (2013) Hybrid Solar-Wind-Diesel Systems for Rural Application in Northern Ethiopia: Case Study for Three Rural Villages using HOMER Simulation. *MEJS* V5: 62-80.
9. Dan Chiras (2010) Wind Power Basics: a Green Energy Guide. New Society Publishers.
10. Olatomiwa L, Mekhilef S, Huda ASN, et al. (2015) Techno-Economic analysis of hybrid PV-diesel-battery and PV-Wind-Diesel-Battery power system for mobile BTS: the way forward for rural development. *Energ Sci Eng* 3: 271-285.
11. Acharya D (2013) Solar and Wind Hybrid Power for an Extremely Remote Mobile Base Station. *Guelph Eng J* 1-10.ISSN: 1916-1107.
12. NASA/SSE. Surface meteorology and energy (2013) Available at http://eosweb.larc.nasa.gov/sse.
13. HOMER, the Micro-Power Optimization Model; ver.2.68Beta, NREL; 2009.
14. Ambekar P, Sengar KP (2015) Cost Analysis of a hybrid system by using An Optimalzation Technique. *Int J Adv Technol Eng Res (IJATER)* 5: 2250-3536.
15. Bhandari B, Lee K-T, Lee G-Y, et al. (2015) Optimization of Hybrid Renewable Energy Power Systems: A Review. *Int J Precis Eng Manuf-Green Tech* 2: 99-112.
16. Phrakonkham S, Le Chenadec J-Y, Diallo D, et al. (2010) Reviews on Micro-Grid Configuration and Dedicated Hybrid System Optimization Software Tools: Application to Laos. *Eng J* 14:15-34.
17. Omari O, Ortjohann E, Mohd A, et al. (2007) An Online Control Strategy for DC Coupled Hybrid Power Systems. *IEEE Power Eng Soc General Meeting,* 1-8.
18. Setiawan AA, Zhao Y, Lee RS, et al. (2009) Design, Economic Analysis and Environmental Considerations of Mini-Grid Hybrid Power System with Reverse Osmosis desalination Plant for Remote Areas. *Renew Energ-Elsevier* 34: 374-383.
19. ARE - shining a Light for a progress, Hybrid power systems based on renewable energies- A suitable and cost-competitive solution for rural electrification. ARE-WG-Technological-Solutions, 200.
20. IFC, GSMA, Green Power for Mobile (2014) Tower Power Africa, Energy Challenges and Opportunist for Mobile Industry in Africa. World Bank Group in partnership with the Netherlands.
21. GSMA, Green Power for Mobile (2014) The Global Telecom Tower ESCO Market Overview of The Global Market For Energy To Telecom Towers In Off-Grid And Bad-Grid Areas. World Bank Group in partnership with the Netherlands.
22. Girma Z (2013) Technical and Economic Assessment of solar PV/diesel Hybrid Power System for Rural School Electrification in Ethiopia. *Int J Renew Energ Res* 3.
23. Lambert T, Gilman P, Lilienthal P (2005) Micro power system modeling with HOMER. *Integration of Alternative Sources of Energy*, F. A. Farret, M.G. Simoes, John Wiley & Sons, 379-416.
24. Pramod Jain (2011) Wind Energy Engineering. McGraw-Hill.

4

Analyses of optimum generation scenarios for sustainable power generation in Ghana

Albert K. Awopone and Ahmed F. Zobaa *

Department of Electronic and Computer Engineering, Brunel University London, Uxbridge UB8 3PH, UK

* **Correspondence:** Email: azobaa@ieee.org

Abstract: This study examines optimum generation scenarios for Ghana from 2010 to 2040. The Open Source Energy Modelling System (OSeMOSYS), an optimisation model for long term energy planning, which is integrated in Long-range Energy Alternatives Planning (LEAP) tool, was applied to model the generation system. The developed model was applied to the case study of the reference scenario (OPT) which examines the least cost development of the system without any shift in policy. Three groups of policy scenario were developed based on the future possible energy policy direction in Ghana: energy emission targets, carbon taxes and transmission and distribution losses improvements. The model was then used to simulate the development of technologies in each scenario up to 2040 and the level of renewable generation examined. Finally, cost benefit analysis of the policy scenarios, as well as their greenhouse gas mitigation potential were also discussed. The results show that: suitable policies for clean power generation have an important role in CO_2 mitigation in Ghana. The introduction of carbon minimisation policies will also promote diversification of the generation mix with higher penetration of renewable energy technologies, thus reducing the overall fossil fuel generation in Ghana. It further indicated that, significant greenhouse emissions savings is achieved with improvement in transmission and distribution losses.

Keywords: generation; Ghana; greenhouse gases (GHG); LEAP; optimisation; renewable energy

---

1. Introduction

Ghana faces serious energy related challenges as the country struggles to meet generation requirement. Inadequate expansion plan of the generation system has resulted in (the current) a situation where 65% of the demand is met as at March 2014 [1]. The electricity supply system of the country is characterised by power outages, which has serious implications on the quality of life as well as industrial development. Reliable and affordable electricity generation is an indispensable commodity in the development of any country [2]. This is particularly important especially with the recent development of the country's oil and gas industry, which has a high potential in attracting investors to an expected oil and gas driven economy.

The generation system of Ghana has relied heavily on hydropower for many years. This high hydropower dependence makes the generation system to be influenced by seasonal cycles. Occasional cases of prolonged dry seasons coupled with insufficient rainfall during the raining seasons have resulted in blackouts and power rationing as experienced in 1984 due to the severe drought conditions of 1983 [3]. Thermal power generation was introduced to supplement the conventional Hydroelectricity after the drought in 1983 underscored the need to diversify the country's generation system. The introduction of Thermal power generation occurred in 1997 with the construction of a combined cycle power plant with an installed capacity of 330 MW at Aboadze near Sekondi-Takoradi. The Takoradi Thermal Power Station (TAPCO), as it is officially called, was eventually expanded to 550 MW with the addition of 2×110 MW combustion turbine plants in 2000. This marked the beginning of a gradual shift to thermal generation in the country. The installed capacity as at March 2014 was 2851 MW which was made up of 1580 MW (55.4%) from the three hydro dams, 1248 MW (43.77%) from Thermal plants and only 2.5 MW (0.09%) from photovoltaic plant [1]. Volta River Authority (VRA), operates all the plants except Bui, which is operated by Bui Power Authority (BPA). VRA and BPA are both government agencies. Sunon-Asogli and Cenit energy are private entities which contribute about 11.61% of the installed capacity [1]. The additions of these thermal plants, however, have not fully solved the energy supply situation, as load shedding has persisted within the last decade [4].

An analysis of the trend in expansion shows that Ghana is gradually shifting to a predominantly fossil-fuelled thermal generation. However, the adverse environmental and societal impacts and fluctuation in the prices of fossil fuels in the world market has necessitated the exploitation of sustainable power generation technologies. These technologies include renewable energy sources such as hydroelectricity, solar energy, wind, wave energy, bio-energy and geothermal energy. Renewable energy sources are naturally replenished at rates that far exceed their consumption. Ghana is endowed with several renewable energy resources, which can be exploited to help meet its energy requirements. There is an excellent solar radiation all year round, and in every part of the country, with an average radiation of about 5 kWh/m^2. Sites suitable for medium and small hydro power plants have also been identified in various part of the country with a potential of adding over 900 MW to the national grid if fully exploited. Sites near the coastal parts of the country have been mapped with excellent conditions for wind generation [5]. The country also has a huge potential for biomass and waste to energy generation [4].

The main challenge with the implementation of renewable energy systems especially in developing countries such as Ghana has been their high initial capital investment. However, it is expected that advancing scientific knowledge will lead to performance improvements and cost reductions of these technologies in the future. The average cost of installing commercial PV systems for instance, fell from about 9.00 US$/watt in 2002 to 5.00 US$/watt in 2012 [6]. This cost reduction is attributed largely to the fast decline in PV modules cost, which reduced by about 85% to 0.8 US$/watt within the same period [6]. Investment decisions and policies to promote renewable energy system will encourage a more rapid development of these generation systems.

The sustainable energy decision-making is often unique and depends on the circumstances of each planning area. Most studies on the generation system of Ghana have only focused on assessing the renewable potential of Ghana. The energy commission of Ghana has projected demand and supply scenarios for Ghana up to 2020, and their report presented as part of the Strategic National Energy Plan of the country [7]. This report was carried out in 2006 and hence does not fully account for current trends. The analysis does not also consider the impact of future policies or the effect of greenhouse gas emissions. Provisions were not also made to identify least cost options for future generation.

The LEAP tool was applied in [8] to analyse the current generation expansion plan of Ghana and to discuss alternative approaches, pointing out the impact on the environment by different generation systems.This current study aims at developing least cost generation pathways for the generation system of Ghana making use of the available technologies and policy drivers as well as challenges in Ghana. Good decisions, concerning the best choice of technologies for future years, will avoid expensive changes in later years. This study will provide a useful insight to guide energy planers and policy makers by providing policy impact assessment of the optimum generation system of Ghana.

2. Materials and Method

2.1. LEAP/OSeMOSYS optimization model

The Long-range Energy Alternatives Planning (LEAP) system, version 2015.0.30.0 was applied to determine the optimal expansion and dispatch of power plants for Ghana from 2010 to 2040. LEAP is a hybrid energy modelling tool which supports a wide variety of methodologies [9]. The optimisation function in LEAP is performed through integration with Open Source Energy Modelling System (OSeMOSYS) which depends on the GNU Linear Programming Kit (GLPK), for solving large linear programming problems [9]. OSeMOSYS is transparent and straight-forward energy modelling tool which allows for simple refinements for sophisticated analysis [10]. OSeMOSYS model compared favourably when validated with established but not freely available MARKAL/TIMES model [10]. LEAP interface with OSeMOSYS provides a transparent connection which enables LEAP to automatically write data files needed for OSeMOSYS. This enables LEAP users to perform optimisation without directly interacting with OSeMOSYS or GLPK [9].

The LEAP model developed by Stockholm Environmental Institute (SEI) is a widely used energy modelling tool for energy policy analysis and Greenhouse gases (GHG) emission mitigation studies, especially in developing countries. The model was adopted for analysis of alternative scenarios and their implications on the electricity sector of Lebanon [11], for examining high levels of renewable energy technologies in Ireland [12] and most recently, for assessing policies aimed at promoting low carbon development in Nigeria [13].

The objective function of the OSeMOSYS model is to estimate the generation system to meet demand by minimising the total discounted cost [14]:

$$[\,Minimise \sum_{y} \sum_{t} \sum_{r} TC_{y,t,r} = OC_{y,t,r} + CC_{y,t,r} + EP_{y,t,r} - SV_{y,t,r}, \forall y,t,r\,] \tag{1}$$

Where, TC_{ytr}, $OC_{y,t,r}$, $CC_{y,t,r}$, $TEP_{y,t,r}$, $SV_{y,t,r}$ represents discounted total cost, operating cost, capital cost, technology emission penalty and salvage value respectively. y, t, r, are the year, technology and region indexes respectively. A full description of LEAP methodology is presented in [9].

2.2. Development of Ghana LEAP model

The Ghana LEAP model was designed with 2010 as the base year, to analyse the possible developmental structure of the generation system of Ghana up to 2040. The choice of this base year was due to availability of data on the national census conducted by the Ghana Statistical Service and a national energy survey by the Ghana Energy Commission which were conducted in 2010, providing reliable data for the model. The selection of 2010 also provided opportunity to validate the results with real data for the past years (2010 to 2015).

The Bottom-Up methodology in LEAP was adopted to model the future energy demand. The population of Ghana in 2010 was 24.7 million people which, was projected to be increasing at a growth rate of 2.4% [15]. In 2010, 83.8% of the urban households had access to electricity. Urbanisation in 2010 was 55.8%, which is expected to increase to about 60% by 2040 [15]. Microeconomic input data was taken from Ghana statistical service [16]. The Gross Domestic Product (GDP) growth rate of 8% was used for base case load projection from 2010 to 2020, increasing to 12% from 2020 to 2040 when the power supply is expected to improve. The energy intensity data used for the model was developed from an energy consumption survey conducted by the Energy Commission of Ghana in 2010 [1]. A real discount rate of 10% was assumed, consistent with West Africa Power Pool generation and transmission master plan recommendations [17].

The dependent installed capacity of Ghana in 2010 was 1865 MW consisting of two large hydro dams with a total capacity of 1040 MW with the remaining contribution from six thermal power plants. To model the generation system, the various generating plants were aggregated. This means that single hydro and thermal plants were modelled to represent the system. The maximum capacity addition was assumed based on Ghana Grid Company (GRIDCO) [18] recommendation to ensure stability of the grid, while the maximum availabilities of the Non-Conventional Renewable Energy Technologies were assumed based on the confirmed available resources [5,7].

The technology cost data was adopted from International Energy Agency [19], National Renewable Energy Laboratory [20] and GRIDCO [18]. The future year investment cost of the conventional energy systems in Ghana (large Hydropower and thermal power) were assumed to be constant throughout the study period while that of renewable systems were assumed to decrease per projections presented in [19]. The prices of fossil fuel resource are particularly very difficult to predict because of its high price fluctuations in the world market. However, the bench-mark fuel price projections in [21] were considered as the most reliable assumptions and hence adopted for this paper.

2.3. Description of scenarios

The scenarios developed are based on cost minimisation with the aim of exploring the potential diversification of the generation system using renewable energy technologies. The scenarios are categorised under four main themes: Reference Scenario (Ref), Carbon Tax scenarios, Energy Efficiency Scenarios and Transmission and Distribution losses scenarios. The policy scenarios are all variations of the Ref scenario under the various constraints described under each scenario. The scenarios represent possible future paths for promoting renewable power generation in Ghana by considering the factors most likely to influence future RET policies in the country. Table 1 show the key elements of the reference and the alternative scenarios.

Table 1. Overview of reference and alternative scenarios.

	Reference	Emission Targets	Carbon tax	T&D
Driving philosophy	Follows continues trend in existing energy policies	Inspired by clean technologies and increasing resistance from residents and environmentalist	Inspired by clean technologies and increasing resistance from environmentalist	Inspired by desire to improve the efficiency of the transmission and distribution system
Demand side	The Bottom-Up demand model approach in LEAP was adopted for modelling the future energy demand. Demand projections to follow official projections [16]. Total demand in 2020 will be 18.88 GWh, increasing to 62.5 GWh by 2040.			
Supply side/ constraints	Least cost development of technologies and dispatch schedule determined by LEAP	Least cost system under emission targets constraints. maximum emission.	Least cost system under emission targets constraints.	Least cost with improvements in transmission and distribution losses.

2.3.1. Reference (Ref) scenario

The Ref scenario is the least cost generation system of Ghana calculated by LEAP/OseMOSYS optimisation model without any policy interventions. LEAP/OSeMOSYS developed the least cost model by deciding both the type and size of technologies as well as the dispatch scheduling. This scenario assumes no shift in policy and serves as the reference for analysing the alternate scenarios.

2.3.2. CO_2 emission reduction target scenarios

Ghana, being a non-annexed I country, has no legally binding emission reduction target in place. Thus, studies evaluating the impact of GHG mitigation are limited in the country. However, the consideration of the impact of GHG by different power generation plants, will lead to a more balanced evaluation of these technologies and will also conform to the Bali road map which underscores the need for developing Non-Annex I countries to introduce mitigation actions after 2012 [22].

The study examines three CO_2 reduction scenarios which were developed making use of the Ref scenario emissions as the benchmark. They are: Low Carbon Emission Reduction (ET), Medium Carbon Emission Reduction scenario (ET1) and High Carbon Emission Reduction scenario (ET2) which impose maximum emission targets of 10%, 20% and 30% respectively, of lower emission targets compared to the optimum scenario. These emission targets were based on the official targets presented Ghana's intended nationally determined contribution (INDC) as part of the country's Intended Nationally Determined Contribution (GH-INDC) presented to the United Nations Framework Convention on Climate Change (UNFCCC) in December 2015 [23].

2.3.3. Carbon tax scenario

The Kyoto protocol makes provisions for three market-based mechanisms (Clean Development Mechanism (CDM); Joint Implementation (JI) and Emission Trading (ET), that provide additional means for countries with commitment to achieve their set targets. These mechanisms have resulted in the creation of the "carbon market" which enables international trading of emission target points [24]. CDM specifically targets emission reduction projects for developing countries. CDM allows Annex I countries to participate in the implementation of emission reduction projects in non-annex I countries. These projects are then assessed as supplementary to domestic projects, to help achieve their targets. Ghana can therefore benefit significantly from these mechanisms to help develop its abundant renewable energy resources, if the cost of domestic GHG emission reduction is lower than the global carbon price.

Another important source of funding for renewable energy development is the introduction of carbon tax. High investment cost of renewable energy technologies (RET) has been cited widely as one of the most important obstacles hindering their development in the developing world including Ghana. Additional revenue from the implementation of carbon tax could be applied directly to the development of RET. Introduction of carbon tax will also provide a fair appraisal for assessing the performance of different generation technologies. Therefore, three carbon tax scenarios where explored in this study. Low Carbon Tax (CT) scenario imposes US\$10/tonne CO_2 introduced in 2020 increasing to US\$20/tonne CO_2 in 2030; Medium Carbon Tax (CT1) imposes US\$30/tonne CO_2 introduced in 2020 increasing to US\$60/tonne in 2030; High Carbon Tax (CT2) imposes US\$50/tonne CO_2 introduced in 2020 increasing to US\$100/tonne in 2030.

2.3.4. Transmission and distribution (T&D) losses scenario

In 2010, transmission and distribution (T&D) losses accounted for about 23% of the total power generation in Ghana, a reduction from a peak of 29% in 2003. However, between 1970 and 2000, these losses varied from 4–11% [25]. The current high T&D losses are due to large non-technical losses (mainly because of theft and non-payment of bills) [7]. The average value of the technical losses (losses due to current flowing through the network) was about 5% between 2010 and 2014 with a long-term projection of 3% [18].

The T&D scenario therefore assumed improvement in the transmission and distribution networks and billing system resulting in reduction in these losses. This was interpolated to reduce to 15% in 2020 and 10% in 2030 in Low Transmission and Distribution scenario (TD). In the Modest Transmission and Distribution scenario (TD1), the T&D losses were interpolated to reduce to 12% in 2020 and 8% in 2030. While in the High Transmission and Distribution scenario (TD2), it was interpolated to reduce to 10% in 2020 dropping further to 5% by 2030.

3. Results and Discussion

3.1. Technical results

3.1.1. Technical results of baseline scenario

The total electricity demand of Ghana increased by about 6 times between 2010 and 2050 (Figure 1). The results of the energy demand projection for Ghana from 2010 to 2040 using LEAP energy demand model is presented in Figure 1. The results show that demand projection will increase into the future with an average demand growth rate of 7% from 2010 to 2040. This growth rate follows the historical demand growth of the country, and are consistent with official load projections [18]

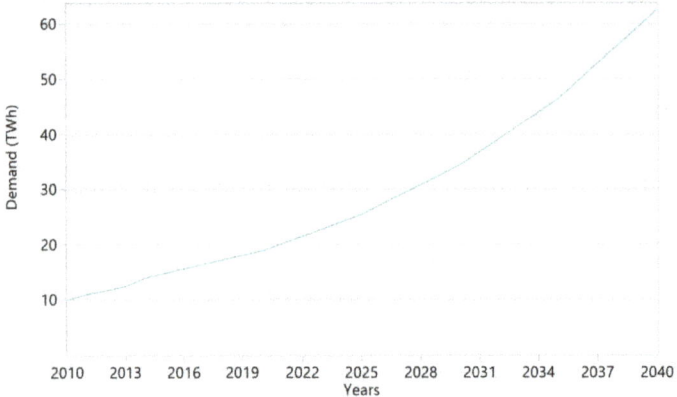

Figure 1. Electricity demand forecast for Ghana.

This implies that without any expansion in capacity, the current generation system will only be able to provide 20% of the demand in 2040. LEAP optimisation algorithm is designed to ensure that supply meets demand requirements. Thus, significant expansion in capacity was predicted in the OPT scenario in order to meet this demand and the 18% reserve margin applied in this study. This value was adopted based on Ghana Grid Company (GRIDCO) [26] recommendation for grid reliability in Ghana.

The Ghana LEAP model predicts a switch from hydropower to thermal generation as illustrated in Figure 2.

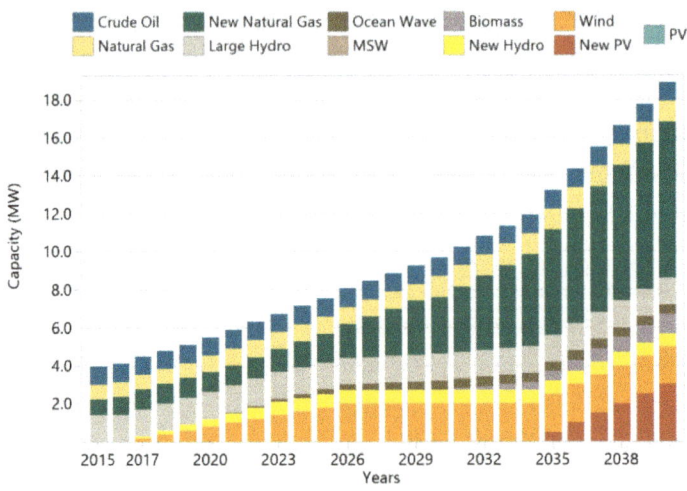

Figure 2. Installed capacity in Ref scenario.

Two main fuel types were considered for the combined cycle thermal generating plants considered in the model. The results show that thermal generation in Ghana will continue to be dominated by natural gas-fuelled systems. Even though the same operating characteristics were modelled for both crude oil and natural gas plants to reflect the co-firing nature of the thermal plants in the country, the model did not consider operating the plants on crude oil as one of the least cost options. This is mainly due to the comparatively higher cost of crude oil. These results give credence to proposals by the Energy Commission Ghana to operate the thermal plants on natural gas when available [7].

Figure 2 also shows the economic competiveness of the Non-Conventional Renewable Energy Technologies (NRET) with the introduction of these systems in much larger quantities from 2016 onwards. The model considered the introduction of wind and small hydro plants in 2016. This means that Wind and Small hydro are the least cost options of the NRET. Wind energy plants capacity increased from 200 MW to its maximum allowable capacity 2025. PV generation plants, which dominated the NRET capacity at the end of the study period, were not considered until 2035, when the model had exhausted the available capacity for wind, hydro and wave generating systems. The model did not however consider Municipal solid waste (MSW) as one of the least cost options for power generation in Ghana due to their extremely high investment cost.

The share of NRET generation in 2010 was 0.059%, which increased to 30% in 2023 because of the introduction of wind and hydro as discussed earlier. This share however, reduced to 25% in 2035, increasing marginally to 27% at the end of the study period with the addition of new PV systems as shown in Figure 3.

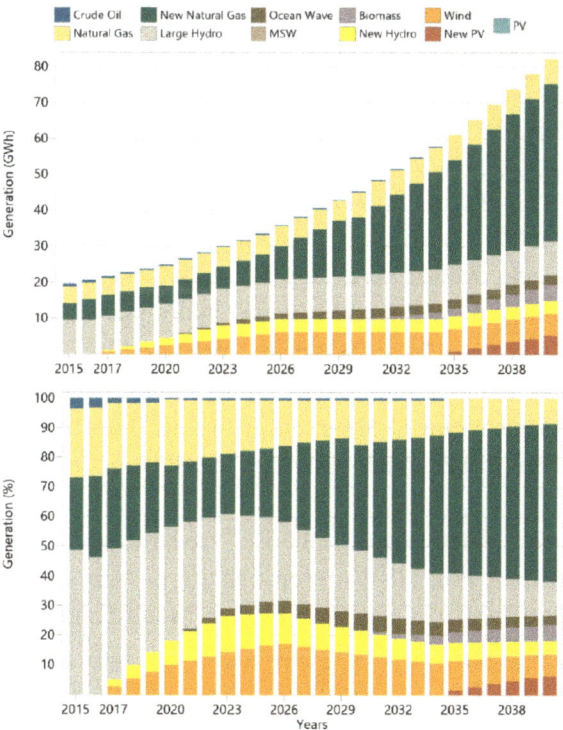

Figure 3. Electricity generation by plant type in Ref scenario.

It is very important to note that the introductions of these NRET were based on their own economic merit. This clearly shows the high potential of diversifying the generation system of Ghana with appropriate renewable energy policies.

3.1.2. Technical results of CO_2 emission target scenarios

The introduction of CO_2 emission targets directly affects the development of clean energy technologies. Figure 4 shows the development of the optimum generation system of Ghana under the three CO_2 emissions constraints. The introduction of these targets resulted in much higher deployment of NRET. Compared to OPT scenario, the model predicted additional 1.5 GW, 3.5 GW and 7 GW NRET generation capacities in the ET, ET1 and ET2 scenarios respectively, by 2040.

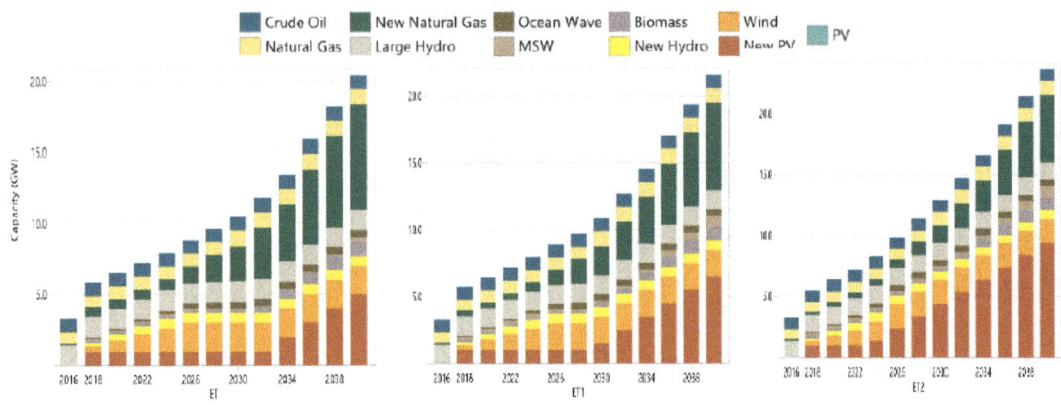

Figure 4. Installed capacity in CO_2 emission target scenarios.

The high NRET have resulted in a total renewable generation capacity share ranging from 54% to 68% with the introduction of 10% to 30% emission reduction targets on the Optimum scenario.

New PV generation plants, which were not considered by the model in the OPT scenario until 2035, are selected at the same time with wind and new hydro plants, with the introduction of carbon emission targets. In the ET scenario, New PV where introduced in 2017 with a capacity of 0.5 GW, increasing to 1 GW in 2018 and remaining constant until 2030, when wind and hydro resources have been fully exploited. The same trend is observed in the ET1 and ET2 scenarios. However, much higher PV capacity addition is achieved with the introduction of high renewable target (ET2 scenario).

MSW plants, which were not selected in the Ref scenario, are forced onto the generation mix with the introduction of carbon tax. MSW generation starts with a capacity of 0.1 GW in 2017, increasing to 0.3 GW in 2040, with a low emission target. Modest and high emission targets will lead to 0.85 GW and 1 GW MSW plants capacity by the end of the study period. The developments of the renewable generation systems have reduced the share of generation from fossil fuel plants. Electricity generation from thermal generation plants in 2040 reduces from 62% in the Ref scenario to between 55% and 42% with the introduction of emission reduction targets as shown in Figure 5.

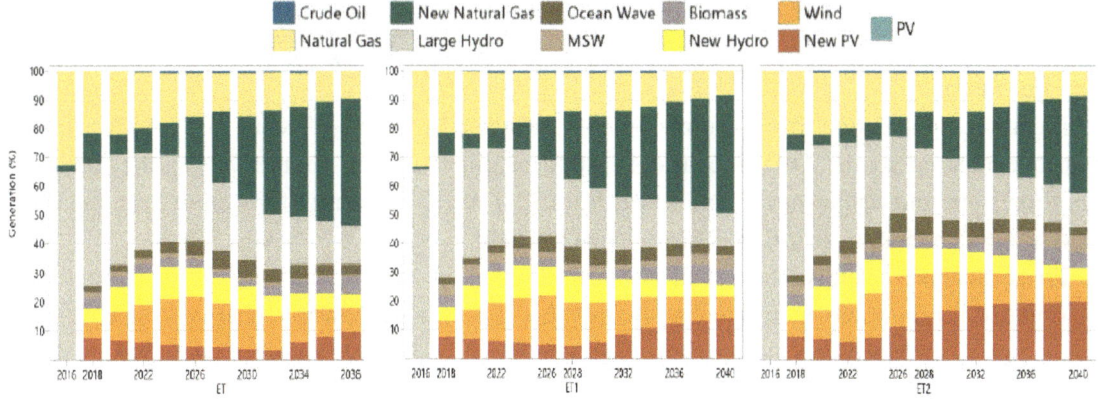

Figure 5. Generation share by plant type in CO_2 emission target scenarios.

This reduction in thermal generation directly reduces the quantity of fuel imports for power generation. This implies that CO_2 targets have a positive impact on energy security of Ghana. The security of the energy system in this case is analysed in terms of self-sufficiency and diversification of energy resources.

3.1.3.   Technical results of carbon tax scenarios

The carbon tax scenarios examine the effect of the introduction of carbon tax constraint on the optimum generation system of Ghana developed by the Ghana LEAP model. The study evaluated three tax levels: CT, CT1 and CT2, representing low, modest, and high carbon tax respectively. Figure 6 shows the development of the optimum generation system from 2015 to 2040 under the three carbon taxes constraints.

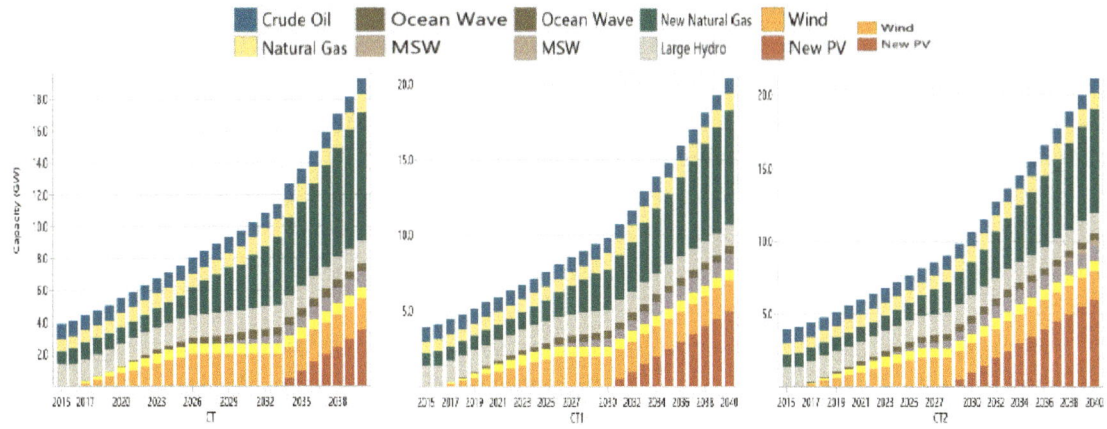

Figure 6. Installed capacity in CO_2 tax scenarios.

It is observed from Figure 6 that the capacity level of the conventional large hydro generation did not change within the study period, however, thermal generation capacity slightly decreases with increase in carbon taxes. The share of thermal plants decreases by about 10% with the introduction high carbon tax (CT2) Scenario. This will result in about 10% savings in fossil fuel requirement for power generation.

Electricity generation from thermal plants reduced by about 17% to 43 TWh in the CT2 scenario compared to the OPT scenario in 2040 as illustrated in Figure 7. On the other hand, generation from renewable sources increased from 32 TWh (38% share of generation) in the BCO to 39 TWh (47% generation share) in CT2 scenario by 2040.

These analyses clearly show that the introduction of carbon tax has direct impact on the development of renewable energy technologies. This is because renewable energy technologies such as wind and PV produce very little to zero GHG emissions. The imposition of carbon tax will affect mainly the cost of production of the thermal plants and thus makes the RET more economically viable.

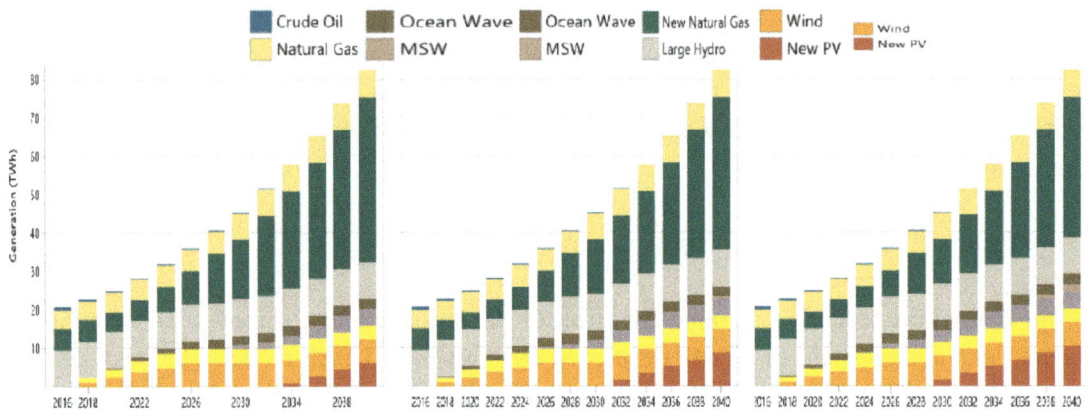

Figure 7. Electricity generation by plant type in CO_2 tax scenarios.

Comparing Figures 2, 4 and 6 shows that both carbon tax and CO_2 emission target scenarios resulted in higher capacities compared to Ref scenario. This is because of their higher renewable energy penetration. The cumulative installed capacities of the CT2 and ET2 scenarios in 2040 are 11% and 26% respectively more than the Ref scenario. These findings are consistent with those of McPherson & Karney [27], which suggest that higher capacities of NRET are required to meet the same demand as thermal plants because of the relatively lower capacity factors of NRET.

3.1.4.   Technical results of transmission and distribution losses scenarios

Transmission and distribution scenarios examine the development of the optimum system with the implementation of strategies to improve transmission and distribution losses. As anticipated, improvement in transmission and distribution losses will lead to lower generation requirements. Figure 8 show the development in generation technologies from 2015 to 2040 with improvement in transmission and distribution losses.

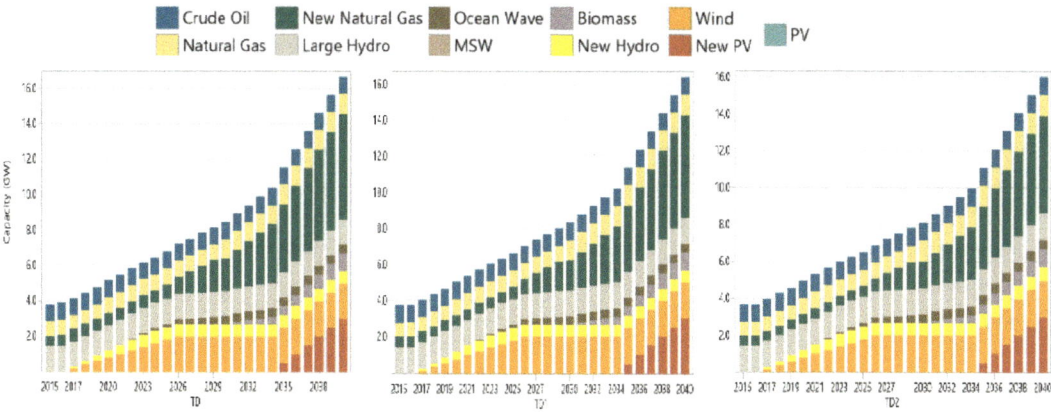

Figure 8. Installed capacity in transmission and distribution losses scenarios.

A reduction of 2000 MW to 3000 MW capacities is achieved depending on the level of transmission and distribution losses improvements, compared to the BCO scenario. This capacity savings is approximately equal to the current (2015) total generation capacity of the country. The development of the generation technologies remains almost the same as that of the BCO scenario. However, much higher renewable generation share of generation is achieve with improvement in transmission and distribution losses.

Ghana electricity generation system has relied greatly on hydropower, and most recently, thermal generation plants. This has resulted in serious power outages and load shedding each time there is insufficient rainfall or inadequate supply of fuel. Introduction of renewable energy policies have resulted in higher diversification of the generation system. The level of energy security of an energy system is improved with higher diversity in energy technologies [14]. These results therefore show that implementation of appropriate renewable strategies will improve the security of the generation system of the country.

3.2. Cost-benefit analysis

The objective of the LEAP/OseMOSYS optimisation is to develop a least cost model of the generation system. Cost-benefits analysis technique was adopted for the economic evaluation of the various scenarios in this study. Cost-benefit analysis is an analytical tool which is used to determine the best possible approach by comparing the cost and benefits of alternatives. In this study, the cost of the system is expressed in terms of Net Present Value of cost of transformation (Investment and Operation and Maintenance cost) as well as resources (fuel cost) over the period of the study discounted at 10% to the base year. Transmission and distribution cost, land use and potential cost of affecting local livelihoods were not considered in this study. This was not only due to lack of reliable data, but also on the scope of this study, which focuses on how the various supply side constraints will influence the development of generation options.

Table 2 presents the results of the cost-benefit analysis provided by the LEAP model. The results compare the cost-benefits of alternative scenarios to Ref scenario. A positive value represents how much more each policy scenario cost compared to the Ref scenario, while a negative number represent benefits.

Table 2. Cumulative costs & benefits: 2010–2040 relative to OPT scenario discounted at 10.0% to year 2010 (billion 2010 USD).

	ET	ET1	ET2	CT	CT1	CT2	TD	TD1	TD2
Transformation	2.2	3	4.3	0.2	0.7	1.2	−0.6	−0.7	−0.8
Resources	−1.7	−2.2	−3.1	−0.2	−0.7	−1.1	−2	−2.4	−2.7
Emission cost				0.3	1	1.5			
NPV	0.5	0.8	1.2	0.3	1	1.6	−2.7	−3	−3.6

It is observed from Table 2 that the cumulative costs of implementing the emission targets and carbon tax scenarios are 0.5 to 1.2 billion USD and 0.3 to 1.6 billion USD respectively, compared to the OPT scenario. This is largely due to the higher capital investment (transformation cost) of RET, which are deployed in much higher capacities. Even though, the constraints on the optimum generation system resulted in savings in the use of resources, these savings were not sufficient to offset the high capital investment.

The introduction of policies to improve transmission and distribution losses to levels proposed in this section will lead to 2.7 to 3.6 billion USD cumulative cost savings over the study period. This significant cost savings is due to the reduction in generation requirement (Figure 7). It should be noted that financial cost in the implementation of measures to reduce transmission and distribution losses were not quantified in this study. This is mainly due to unavailability of reliable information on the topic in Ghana. However, the high savings suggest the need for the country to adopt strategies to reduce these losses. Greater premium should be placed on reducing the non-technical losses, which constitute the bulk of the transmission and distribution losses in Ghana [1]. According to [28], only 50% of the electricity generated in sub-Sahara Africa is paid for. This is attributed to a combination of low percentage of billing electricity injected in distribution networks and low collection of billed amounts. The country should therefore consider the implementation of Advanced metering infrastructure (AMI) for monitoring consumers, the use of temper proof meters and replacement of old meters with accurate electronic meters and frequent energy audits up to the distribution transformer [29]. These measures have proven to be very effective in reducing non-technical losses in North Delhi from 53% in 2002 to 15% in 2009 [29].

3.3. Environmental analysis

The LEAP model contains all Global Warming Potential (GWP) factors, required for environmental assessment. The Forth Assessment Report (AR4) factors were applied in this study in line with Intergovernmental Panel on Climate Change (IPCC) recommendations [30]. Table 3 shows cumulative GHG emissions at points of emissions in the various scenarios.

Table 3 confirms the mitigation potential of the policies evaluated in this study. Significant GHG emission reduction potential of 16–37% is achieved with the introduction of CO_2 emission targets. This is because of the relatively higher share renewable generation. The emission targets compelled higher deployment of renewable energy technologies, since they emit low to zero emissions.

Table 3. Cumulative emissions and cost of avoided CO_2 emissions.

Scenario	Cumulative emission (Mt CO_2e)	CO_2e savings (%)	Cost of avoiding GHG (($/tonne CO_2e)
Ref	223		
ET	187	16	13.9
ET1	170	24	15.1
ET2	141	37	14.4
CT	217	3	53.8
CT1	202	9	48.1
CT2	190	15	48.5
TD	168	25	−48.3
TD1	160	28	−49.3
TD2	150	33	−49.1

Table 3 further reveals that introduction of carbon tax at levels proposed in this study will not yield significant GHG emission savings. The cumulative GHG emission savings in the high carbon tax scenario (CT2) compared to OPT scenario was 15%, while that of ET2 and TD2 scenarios were 37% and 33% respectively. These results show that even though costing emissions will increase the operational cost of thermal generation, it does not automatically lead to adoption of renewable energy technologies. There is a tendency by utility providers to transfer this extra cost to consumers. This negative impact of carbon tax can be address with the investment of tax revenue in the development of renewable technologies.

The cost effectiveness of GHG emission was analysed by calculating the cost of avoiding CO_2 emission. Cost of CO_2 avoided is the cost of reducing CO_2 emission to the atmosphere expressed as $/tonne of CO_2 not emitted with respect to the base case scenario. The decision criterion is to identify the least cost alternative in reducing a tonne of CO_2. The results show superior performance of the transmission and distribution losses scenarios in achieving economically efficient CO_2 emission reduction. While all the other policy scenarios resulted in abatement cost, the transmission and distribution losses scenarios resulted in benefits. This is because of the relatively lower demand achieved with the implementation of policies to reduce transmission and distribution losses. The country can therefore benefit immensely from the global market with the implementation of appropriate energy policies.

4. Conclusion

In this study, LEAP energy tool was used to develop a model of the generation system of Ghana (Ghana LEAP) based on the operational characteristics of the generation of Ghana in 2010. The LEAP model of Ghana was used to determine the optimum development of the country's generation system up to 2040. Three policy constraints were applied to determine their effect on the optimum generation system.

The analyses show that suitable policies for clean power generation have an important role in CO_2 mitigation in Ghana. This was confirmed with the introduction of non-conventional renewable technologies on their own merit in the OPT scenario. Further analysis show that introduction of carbon minimisation policies will promote diversification of the generation mix with higher penetration of renewable energy technologies. This will reduce the overall fossil fuel generation in

Ghana, which is characterised by unreliable feedstock fuel supply, as well as price shocks, leading to improved energy reliability.

The results further show that significant greenhouse emissions savings is achieved with improvement in transmission and distribution losses resulting in net benefits in cost of avoided emissions compared to the base case. Ghana being a signatory to the United Nations Framework Convention on Climate Change (UNFCCC) has an obligation to report periodically on measures taken to reduce greenhouse gases. The analysis of the possible least cost generation pathways show that further reduction in GHG is possible in Ghana with the implementation of additional policies. The study therefore recommends the implementation of strategies to reduce transmission and distribution losses as well as carbon minimisation policies such as emission targets and carbon tax. These measures will promote the development of more efficient, reliable and environmentally acceptable energy system in Ghana.

Further studies need to be carried out to assess the impact of high penetration of renewable energy generation technologies on the stability of the grid as well as grid expansion studies to accommodate the potential future generation expansion.

Conflict of Interest

All authors declare no conflicts of interest in this paper.

References

1. Energy Commission Ghana, Energy (Supply and Demand) outlook for Ghana, 2015. Available from: http://www.energycom.gov.gh/index.php/data-center/energy-outlook-for-ghana.
2. Benefoh DT, Ackom EK (2016) Energy and low carbon development efforts in Ghana: institutional arrangements, initiatives, challenges and the way forward. *AIMS Energ* 4: 481–503.
3. Eshun ME, Amoako-Tuffour J (2016) A review of the trends in Ghana's power sector. *Energ Sustain Soc* 6: 1–9.
4. Asumadu-Sarkodie S, Owusu PA (2016) A review of Ghana's solar energy potential. *AIMS Energ* 4: 481–503.
5. Gyamfi S, Modjinou M, Djordjevic S (2015) Improving electricity supply security in Ghana: the potential of renewable energy. *Renew Sustain Energ Rev* 43: 1035–1045.
6. Fieldman D (2014) Photovoltaic (PV) pricing trends: historical, recent, and near-term projections. Available from: http://escholarship.org/uc/item/06b4h95q.pdf.
7. Energy Commission Ghana, Strategic national energy plan 2006–2020, 2006. Available from: www.energycom.gov.gh/files/snep/MAIN%20REPORT%20final%20PD.pdf.
8. Awopone AK, Zobaa AF, Banuenumah W (2017) Techno-economic and environmental analysis of power generation expansion plan of Ghana. *Energ Policy* 104: 13–22.
9. Heaps CG long-range energy alternatives planning (LEAP) system, 2016. Available from: https://www.energycommunity.org.
10. Howells M, Rogner H, Strachan N, et al. (2011) OSeMOSYS: the open source energy modeling system: an introduction to its ethos, structure and development. *Energ Policy* 39: 5850–5870.
11. Dagher L, Ruble I (2011) Modeling Lebanon's electricity sector: alternative scenarios and their implications. *Energy* 36: 4315–4326.

12. Welsch M, Deane P, Howells M, et al. (2014) Incorporating flexibility requirements into long-term energy system models: a case study on high levels of renewable electricity penetration in Ireland. *Appl Energ* 135: 600–615.

13. Emodi NV, Emodi CC, Murthy GP, et al. (2017) Energy policy for low carbon development in Nigeria: a LEAP model application. *Renew Sustain Energ Rev* 68: 247–261.

14. Augutis J, Martišauskas L, Krikštolaitis R (2015) Energy mix optimization from an energy security perspective. *Energ Convers Manage* 90: 300–314.

15. Ghana statistical service, 2010 population and housing census-national analytical report, 2013. Available from: http://www.statsghana.gov.gh/publications.html.

16. Ghana statistical service, Revised 2015 annual gross domestic product. Available from: http://www.statsghana.gov.gh/gdp_revised.html.

17. Update of the ECOWAS revised master plan for the generation and transmission of electrical energy, 2011. Available from: http://www.ecowapp.org/?page_id=136

18. Generation system master plan of Ghana, 2011. Available from: http://www.gridcogh.com/en/press-media/electricity-supply-plan.php.

19. Energy technology perspectives 2014, 2014. Available from: https://www.iea.org/etp/etp2014/.

20. Cost and performance data for power generation technologies, 2012. Available from: https://www.iea.org/etp/etp2014/.

21. Annual energy outlook 2015 with projections to 2040, 2015. Available from: www.eia.gov/forecasts/aeo.

22. Yi J, Zhao D, Hu X, et al. (2016) An integrated CO_2 tax and subsidy policy for low carbon electricity in Guangdong, China. *Energ Source* 11: 44–50.

23. Ghana's intended nationally determined contribution (INDC) and accompanying explanatory note, 2015.Available from: http://www.theroadthroughparis.org/negotiation-issues/indcs-submitted-date-0.

24. UN Climate Change Room, 2016. Available from: https://www.epa.gov/ghgemissions/sources-greenhouse-gas-emissions.

25. Electric power transmission and distribution losses (% of output), 2014. Available from: http://data.worldbank.org/indicator/EG.ELC.LOSS.ZS.

26. Ghana Wholesale power reliability assessment, 2010. Available from: http://www.ecowrex.org/document/ghana-wholesale-power-reability-assessment.

27. McPherson M, Karney B (2014) Long-term scenario alternatives and their implications: LEAP model application of panama's electricity sector. *Energ policy* 68: 146–157.

28. Antmann P (2009) Reducing technical and non-technical losses in the power sector. Available from:http://documents.worldbank.org/curated/en/829751468326689826/pdf/926390WP0Box380 0in0the0power0sector.pdf.

29. Agüero JR (2012) Improving the efficiency of power distribution systems through technical and non-technical losses reduction. *IEEE PES*, 1–8.

30. Report of the conference of the parties on its nineteenth session, held in warsaw from 11 to 23 November 2013, 2014. Available from: http://unfccc.int/resource/docs/2013/cop19/eng/10a03.pdf.

5

Energy security and competition over energy resources in Iran and Caucasus region

Shahrouz Abolhosseini [1], Almas Heshmati [2,*], and Masoomeh Rashidghalam [3]

[1] National Iranian Oil Company, International Affairs, Petroleum University of Technology, Teheran, Iran
[2] Sogang University, Seoul, Korea
[3] University of Tabriz, Tabriz, Iran

* **Correspondence:** Email: almas.heshmati@gmail.com.

Abstract: Energy security as a dominant factor in international stability is of great importance for major economies. The global energy market with its current level of supply and demand relies on energy sources in the Middle East, Caucasus, Central Asia and Russia. After the Fukushima disaster nuclear powers in Europe view renewable energy sources as a serious alternative. Europe's energy vulnerability has deteriorated due to the Russia-Ukraine conflict. However, renewable energy sources are not large enough to replace nuclear power completely. This trend will continue with climbing demand especially in the natural gas sector as clean energy. In this research, Caucasus and Iran are considered the main sources and routes for energy transmission to the global market, including Europe. Caucasus plays a key role in bridging Europe and Asia. Also, Iran is an alternative for energy transmission to Europe after lifted sanctions. As part of the European active supply diversification policy Iran has capacity to reduce Europe's energy dependency on Russia. However, changes in US new administration America First Policy is harmful for the EUs energy security. Caucasus aims to catch a large share of the European energy market since the Baku-Tbilisi-Ceyhan pipeline has started operations and Iran is also trying to expand its market to become a sustainable source of energy for major consumers. Therefore, Iran and Caucasus are considered reliable energy suppliers for Europe. In this regard, we analyze the best motivation for changing the direction new suppliers' energy policies towards Europe and suggest alternative solutions to compete with rival countries in order to enhance energy security.

Keywords: energy resources; energy security; oil and gas; Caucasus; Iran

Abbreviations:

b/d	barrels per day
bcm	billion cubic meters
BTC	Baku-Tbilisi-Ceyhan
BTE	Baku-Tbilisi-Erzrum
CO_2	Carbon dioxide
CPC	Caspian Pipeline Consortium
EC	European Community
EU	European Union
HHI	Herfindahl-Hirschman Index
JCPOA	Joint Comprehensive Plan of Action
LPG	liquefied Petroleum Gas
MoUs	Memorandums of understanding
mn t/yr	million tons per year
Mtoe	Million tons oil equivalence
NATO	North Atlantic Treaty Organization
OPEC	Organization of the Petroleum Exporting Countries
RSC	Regional Security Complex
UK	United Kingdom
UN	United Nations
USA	United States of America

1. Introduction

Energy, as a fundamental component of modern society and life, has a direct impact on human activity and it plays a critical role in socioeconomic development. Energy is deeply embedded in each component of mankind's development [1–4]. Reliable and sustainable energy supply is therefore a fundamental condition for economic development and growth. It has also become essential for the smooth functioning of modern economies [5].

Increasing global energy demand, concerns about energy security such as availability of fossil fuels and the dependency on them, anthropogenic emissions of greenhouse gases and environmental degradation caused by energy generation from fossil fuels has stimulated debates about the future efficacy of fossil fuels [6–9]. A legal definition of energy security is still in the process of being formulated and others [10] have contributed to this definition by stating it as a condition in which a nation and all, or most of its citizens and businesses have access to sufficient energy resources at reasonable prices for the foreseeable future free from serious risks of a major disruption of services.

There are two new issues which put secure energy supply into question. First, global climate change has already started and its catastrophic consequences are increasing. The main reason for this is the growing consumption of fossil fuels. Second, the profound transformation of the energy market–the consequences of severe growth in demand, the state-monopolistic approach of countries with raw materials, shareholder value oriented strategies of private energy companies and the looming exhaustion of oil reserves–has given rise to a situation in which satisfying demand is a matter of constant anxiety. On the other hand, the remaining oil and gas reserves are concentrated in politically unstable regions. The renewed focus on energy security is driven because of exceedingly tight oil markets and high oil prices. But it is also fueled by the threat of terrorism, instability in some exporting countries, geopolitical rivalries and countries' fundamental need for energy to increase their economic growth.

However, the uneven distribution of energy supplies among countries has led to significant vulnerabilities [5]. According to the International Energy Agency, 95% of the global economy is affected by the decisions of 5 or 6 states in the Middle East which are facing ethnic problems, political crises, terrorism, corruption and authoritarianism [11]. Among others, Saudi Arabia and Iran play a major role in the energy security of the world [12]. With this background there is renewed anxiety over whether there will be sufficient resources to meet the world's energy requirements in the decades ahead.

Energy security as a dominant factor of international stability is of great importance for major economies. Changes in energy policy in Europe in the aftermath of the Fukushima disaster and Europe's deteriorating energy vulnerability situation due to the Russia-Ukraine conflict have led to continuous increase in demand especially in the natural gas sector as clean energy. In this research, Caucasus and Iran are considered the main sources and routes for energy transmission to the global market, including Europe. Caucasus aims to catch a large share of the European energy market through its Baku-Tbilisi-Ceyhan pipeline. Iran is also trying to expand its energy supply market. Therefore, Iran and Caucasus are considered reliable energy suppliers for Europe. This research analyzes the motivation for changing the direction of the new suppliers' energy policies towards Europe and suggests ways to compete with rival suppliers in order to enhance market shares and improve general energy security. Examining Joint Comprehensive Plan of Action (JCPOA) and its possible effect on regional energy security and computation of energy security index for selected economies in EU for the period 1980–2014 could be considered as our contribution compared to previous literature. Due to conflict between Iran and Western powers, previous researchers didn't consider Iran as an important player in the region and its role in the regional energy security. Also, there is no analysis regarding the effect of JCPOA implementation on European energy security.

The rest of this study is organized as follows. Section 2 discusses the importance of Iran in energy security. Section 3 gives details of energy supply and demand. The issues of energy security as a driver of alternative energy sources with a focus on the Iran and Caucasus region are discussed in Section 4 while Section 5 looks at energy resources. Section 6 discusses Iran's isolation and its implication for regional security. The final section provides a conclusion.

2. Iran's Importance in Energy Security

This section focusses on literature which has studied Iran's importance in the energy security of the world. Iran's potential contribution to Europe's energy security with emphasis on energy cooperation between Tehran and Brussels is assessed [13]. According to his study progress in energy cooperation between Iran and the European Union (EU) is likely to remain conditioned on reaching a compromise on allegations regarding Iran's nuclear program. Europe's reluctance to invest in Iran's hydrocarbon sector has provided great commercial opportunities for companies from other countries to make lucrative profits. According to the author, European (and American) technology is needed to fully utilize Iran's oil and gas deposits. Therefore, the development of these reserves will contribute to European and global energy security. This will also facilitate Europe's efforts to diversify suppliers and thereby reduce its heavy dependence on Russia.

Iran's role in energy security at the regional and global levels is studied [14]. According to her, under pressure from the United States which aims at the global isolation of Iran, the country has turned its eyes to Europe and China to meet its energy demands and has become a fruitful partner for Russia. Several factors such as neglecting Iran's legal rights in the Caspian Sea and lack of attention to the economic and political advantages of energy transmission through the country favor Russian policies. According to this study it is obvious that due to an idealist approach in its foreign policy Iran has lost the opportunity to play an effective and constructive role in energy security at the regional and global levels. The study suggests that Iran's unique geographical location between the Caspian Sea and the Persian Gulf is still usable and will serve Iran diplomatically. Also, despite the fact that opportunities were not seized, the ongoing regional and global developments continue to bring new opportunities. These opportunities can be exploited by realistically evaluating the situation on the basis of national interests which benefit supplier and consumer nations.

There is a need for infrastructure which will make it possible to supply an unlimited amount of energy from the South Caspian/Persian Gulf region to Europe, which is not completely available [15]. Why isn't there a major pipeline from the South Caspian region to Europe which could be fed from all South Caspian and Persian Gulf states? This study gives political issues as the main reason for this. Private investors will not risk building a pipeline which might be disputed due to US sanctions against Iran. Also America's attitude towards Iran is especially frigid because of Iran's nuclear project and because Iran is allegedly destabilizing the situation in Iraq and beyond. The United States prevents Tehran and individual European investors' plans to transit Iranian energy resources via Turkey to the EU. However, Russia, alongside with the Caspian Sea Region gas suppliers and the Middle East account for about 90% of total global natural gas reservoir [16]. Therefore, the critical role of Iran and Caucasus Region cannot be ignored by Europe.

3. Energy Supply and Demand in Europe

In the last two decades Europe's dependency on natural gas has significantly increased as natural gas has become a main source of energy. This dependency on natural gas emerged as a result of the EU's intergovernmental restrictions on CO_2 emissions, high emissions from coal-based generators and various obstacles in the rapid development of renewable energy [17]. In 2009, about 26% of the primary energy consumption of all 27 EU member states was of natural gas. With an annual growth rate of 2.7%, the total gas demand is projected to rise to 43% of primary energy consumption by 2030 [17].

3.1. Possible alternative energy sources for Europe

Currently, about 50% of the European natural gas is imported by pipeline from outside the continent where the dominant supplier is Russia [16]. Also, 32% of traded crude oil in Europe was sourced from Russia from 2015. European policymakers are trying to reduce their dependency on energy imports by enhancing renewable energy, developing energy efficiency and energy technologies. It is forecast that renewable energy production in Europe and Eurasia will increase to 330.7 Mtoe in 2035 compared to 115.5 Mtoe in 2013 [18]. However, this trend is declining for oil production as estimates show that oil production levels will decrease to 803.3 Mtoe in 2035 as compared to 834.8 Mtoe in 2013 [19].

EU receives energy supplies from a number of countries around the world such as Norway, Russian, Central Asia, Caucasus and the Organization of Petroleum Exporting Countries (OPEC). EU works actively with these countries to get the best deals possible to increasingly diversify its energy sources and to prevent disruptions in supply. In future, additional energy demand from China and India should be met in competition by these sources as well.

Norway is the world's third largest exporter of oil and gas after Saudi Arabia and Russia. In 2012, it accounted for about 31% of all EU's natural gas imports and 11% of its crude oil imports. Another energy source for EU is Russia as more than 40% of Europe's imported natural gas through pipelines comes from Russia [18]. Ukraine and Belarus' gas supply dependency on Russia is 74% and 100% respectively.

A number of individual EU countries are heavily dependent on Russia particularly for natural gas (European Commission)[1]. This dependency leaves them vulnerable to supply disruptions, whether caused by political or commercial disputes and infrastructure failures. For instance, a 2009 gas dispute between Russia and the transit country of Ukraine, left many EU countries with severe shortages[2]. The Russia-Ukraine gas crisis not only illustrated to European states how strong their dependency is on Russian gas, but also emphasized the need to diversify their main energy supply sources as the flow of gas from Moscow can be unstable [20].

3.2. Importance of Iran and Caucasus for Europe energy security

Iran has the highest natural gas reserves in the world (34 trillion cubic meters of natural gas). This amount is sufficient to satisfy current EU natural gas demand for 90 years. Despite this rich natural endowment, the country has not yet translated this potential into reality. Paradoxically, its natural gas production continues to be barely sufficient to satisfy its domestic demand. The Iranian natural gas industry most likely will remain focused on the domestic market, and on limited amounts of export within the region. In the aftermath of the nuclear deal, Iran is set to concentrate its energy strategy on the development of the oil sector. In this framework, more natural gas might be utilized for reinjection into oil fields in order to sustain growing oil production and exports. Iran will try to use its natural gas resources to improve the competitiveness of its economy, through a larger share of power generation based on cheap natural gas, and through further investments in natural gas-fueled vehicles, in a move to reduce the domestic consumption of oil, which could thus be freed-up for exports[3].

Iran had a successful 2016 to capture a share of the European energy market. The bulk of international sanctions targeting Iran were lifted by JCPOA in January 2016 and the country's crude export to Europe has soared to 497,323 bbl/day during 2016 compared to 111,880 bbl/day during the sanction period (1st July 2012 to 31st December 2015). Table 1 shows the crude Iran's oil exported to Europe by destination. These figures could be considered as an indication for supply diversification made by European countries after JCPOA implementation on 16th January 2016.

---

[1] https://ec.europa.eu/energy/en/topics/imports-and-secure-supplies/supplier-countries.

[2] European Commission, https://ec.europa.eu/energy/en/topics/imports-and-secure-supplies/supplier-countries.

[3] http://bruegel.org/2015/10/iran-a-new-natural-gas-supplier-for-europe/

Table 1. Crude oil exported to Europe by Iran after JCPOA (bbl/d).

	Sanction period, 2012–2015	Post sanction period 2016
Austria	-	8,307
Bulgaria	-	1,617
France	-	138,224
Greece	-	56,880
Hungary	-	2,725
Italy	2,569	53,413
Netherlands	-	8,690
Poland	-	8,344
Spain	-	55,522
Romania	-	16,427
Turkey	109,311	138,699
UK	-	8,473
Total	111,880	497,323

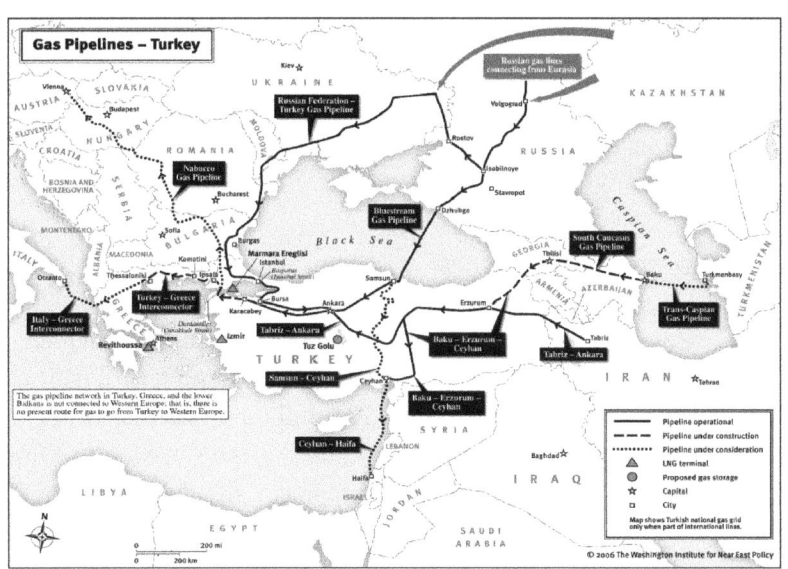

Figure 1. Alternative energy transmission routes for Europe [21].

Countries in Central Asia and Caucasus are rich in natural resources including oil and gas that can help EU diversify its energy supplies. Recognizing this potential EU has been participating in the development of their energy sectors. It has signed memorandums of understanding (MoUs) with Azerbaijan, Kazakhstan, Turkmenistan and Uzbekistan. These MoUs outline steps for further energy cooperation with these countries. The 12 countries that make up OPEC also provide EU with around 40% of its total crude oil imports. Of all OPEC countries, Saudi Arabia, Libya and Nigeria are the biggest individual suppliers, each having delivered over 8% of the EU's total oil imports in 2012. EU meets OPEC annually at the ministerial level to discuss a variety of issues including the promotion of more stable oil prices and transparent markets.

In response to energy security concerns, the European Commission (EC) released its Energy Security Strategy in May 2014. The strategy aims to ensure a stable and abundant supply of energy for European citizens and their economy (European Commission)[4]. The energy supply situation in Europe has deteriorated due to the Russia-Ukraine conflict. Therefore, Iran and Caucasus are considered reliable alternative energy suppliers for Europe.

Figure 1 illustrates the importance of Iran and the Caucasus region as the main alternative sources of energy for Europe.

4.   Energy Security: A Driver for Alternative Energy Sources

Energy security is affected by different factors including high oil prices, dependency on oil imports, depletion of oil fields, political instability in major oil exporting countries, disputes between transition countries and exporters and any disruption in energy supply. The level of security is estimated by the risk of supply disruptions or the costs incurred due to lack of security. The oil market is not considered a competitive market because more than 70% of the total crude oil reserves are located in OPEC [16]. There are some geopolitical regions around the world which are energy sources. The most important ones include the Persian Gulf and the Caucasus region. There are many exporters in these regions which are sustainable sources of energy as they have huge sources of oil and gas. But some of them have suffered from political instability and military conflicts. Caucasus and Iran are considered as major sources of energy and alternative routes for energy transmission to the global market, especially to Europe.

4.1. Energy security in Europe's major economies

Germany, Italy, UK and France are among the European countries which are members of G-7 and are considered major economics. They were also the largest importers of natural gas in Europe in 2014–2015. They imported 104.0, 50.2, 29.0 and 35.9 billion cubic meters (bcm) respectively of natural gas in 2015. Russia is the main source of natural gas transmission to Europe. We assessed energy security as a key factor of sustainable development for these countries by applying the Herfindahl-Hirschman Index (HHI) to measure the level of energy security. We calculated HHI of fuel-mix concentration for Germany, Italy, UK and France for 2000–2015. The results are given in Figure 2.

Our findings show that the selected countries had HHI above 2000, indicating that these countries were highly concentrated in the use of fossil fuels as part of their primary energy consumption.

We also derived a security index to measure the energy security index for some EU countries (see Table 2). This index was estimated by the World Bank based on net energy imports (as percentage of energy use) that are estimated as total energy use less domestic energy production [22]. Table 2 shows that in particular Germany, Spain and Italy were highly dependent on importing energy. It should be noted that UK was a net exporter in 1990 and 2000, but this situation changed in 2010.

---

[4] https://ec.europa.eu/energy/en/topics/imports-and-secure-supplies/supplier-countries.

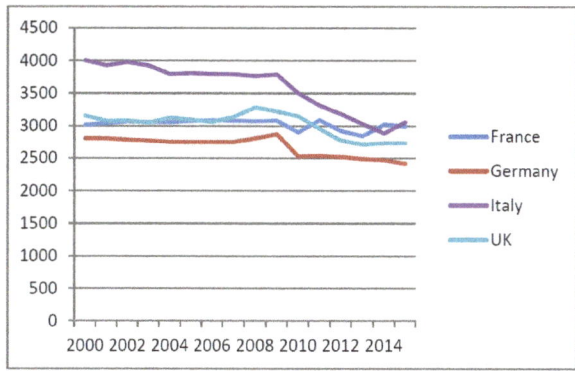

Figure 2. The Herfindahl-Hirschman Index of energy security for major economies in Europe during 2000–2015 [16].

Table 2. Energy security index of selected economies in EU (1980–2014) [22].

Country	1980	1990	2000	2010	2014
Germany	48.03	46.99	59.82	60.67	60.89
UK	29.41	−1.01	−22.21	26.7	39.6
Spain	76.7	61.6	74.1	73.05	69.96
France	72.57	50.06	48.13	48.07	43.58
Italy	84.79	82.73	83.58	82.65	75.38
Sweden	60.16	37.11	35.82	35	29.27

High dependency on energy imports suggests that policymakers in EU have to find alternative sources of energy to replace Russia with more reliable supplier countries. Energy security or security of supply is related to three dimensions: availability, affordability and sustainability. Iran and the Caucasus region have great potential to serve as alternative sources of energy for Europe. Considering the conflicts in Caucasus and the dispute between Iran and the western powers balanced energy security for all countries in the region is required. If there is lack of security for a country compared to the countries in its neighbourhood we cannot expect any cooperation. Any political instability, competition over energy resources or a dominant position in the region may lead to an unstable situation. If there is security cooperation, it should be applied to all parties. Otherwise an unbalanced situation will emerge. International cooperation from all parties is required to convert an Energy Security Complex to an Energy Security Community.

4.2. Caucasus, strategic crossroads

Caucasus is a valuable energy enhancing region due to its location at the crossroads between Europe and Asia. The region also possesses an important supply of energy [23]. Integration of this region into the world market is significant for economic growth in these countries [24]. In this context, Azerbaijan, Georgia and Turkey constructed energy transportation routes in Caucasus in the early 1990s to link these countries to Europe and Asia [25]. In particular, the development of the Baku-Tbilisi-Ceyhan (BTC) oil pipeline and the Baku-Tbilisi-Erzurum (BTE) link Azerbaijan,

Georgia and Turkey to Europe and Asia [26]. It should be noted that Turkey and Georgia are only transit routes while Azerbaijan and the countries located on the east side of the Caspian Sea like Turkmenistan, Kazakhstan and Uzbekistan are energy suppliers.

The BTC pipeline transits oil from Baku in Azerbaijan to Tbilisi in Georgia and Ceyhan in Turkey. Its construction started in 1998 and the pipeline was completed in 2005 [27]. The US had been pushing hard for BTC as the first pipeline specifically designed to export Caspian oil out of the region without going through Russia. It can transport 1 million barrels of oil a day from Azerbaijan via Georgia to the Turkish port of Ceyhan. Alongside BTC runs the BTE (or South Caucasus) gas pipeline through which Turkey imports gas from Azerbaijan. The BTE pipeline is used for transferring natural gas from Baku in Azerbaijan to Tbilisi in Georgia and further to Erzurum in Turkey. The construction of the BTE gas pipeline started in 2004 and was finished in 2007 [26].

The Caspian Pipeline Consortium (CPC) is a consortium and a major export pipeline to transport crude oil from the Kashagan and Karachaganak fields in Kazakhstan to the Novorossiysk on Russia's Black Sea coast. CPC will complete the expansion plan of this crude export route and raise the capacity to 67 mn t/yr by the end of 2017 from 52 mn t/yr [28]. The CPC route is expected to transport 65.7 mn t/yr of crude this year of which more than 80% is sourced from Kazakhstan and the balance is supplied by the Russia. The most important parameter in the geopolitics of Caspian Region and Caucasus is control of oil and gas transition. Energy transportation through BTC, CPC or any possible pipeline would affect the region from different point of views including; energy perse (China), economic implications (Turkey, Iran, and the oil companies), a way to gain influence and/or prevent others from doing so (the USA and Russia) [29].

Long-distance, cross-border pipelines are important to expand energy security and make an alternative to the many vulnerable checkpoints along the sea transportation routes [30]. However, it is important to note that the construction of energy transportation routes has circumvented Armenia and Russia, which in turn, has shaped the economic, social and political integration in the Caucasus region [31]. In recent years, independent of the central Government of Iraq, the Kurdistan region of the country has completed the construction of pipelines linking the region's oil and gas fields to Ceyhan in Turkey and Iranian energy supply routes with daily exports of more than 500,000 barrels of oil [32].

4.3. Iran, a geopolitical energy supplier

Another alternative route for energy transmission to Europe with capacity to significantly reduce its energy dependency on Russia is Iran. The Russian Federation, which performs as a global player, is accused of taking advantage of the dependency of other countries by cutting off energy resource exports in order to reach its political aims. Also, the recent energy crisis between Russia and Ukraine has threatened European energy supply. Therefore, it has forced European countries to double their efforts to diversify their energy resources and energy transportation routes. Developments in the aftermath of the collapse of the Soviet Union along with the independence of its former republics in Caucasus and Central Asia have hastened the importance of Iran in the northern part of Hormuz Strait in the Persian Gulf [15].

On the other hand, Iran is located between two strategic energy reserves of the world-Caspian Sea and the Persian Gulf, and it enjoys a unique situation. Besides Iran also borders Central Asian countries through its neighbor Turkmenistan and it has also become a fruitful regional partner for the

Russian government in the post-Soviet Union era [33]. According to an announcement by European politicians Iran's role in energy supply to Europe is becoming more important [15].

Iran is the world's third and OPEC's second largest oil exporter with a share of 15% in the Persian Gulf and 6% in the world oil market. Iran exports about 3,000,000 barrels of petroleum products per day, which is 1.3% of the total world exports. Iran exports liquefied petroleum gas (LPG) and distillate fuel oil. Iran is also the main importer of petroleum products in the Persian Gulf region. The country is ranked among the top countries that have many natural resources such as crude oil and natural gas. Its oil and gas are allocated 44% and 54% of the total energy share respectively [34]. For over ten years Iran produced approximately 3.5 million barrels of oil per day (b/d) (currently 5% of the world production) with a rather narrow variance of less than 0.3 million b/d. Slightly more than 2 million b/d of production is available for exports. Iran is the fourth largest oil producer (behind Saudi Arabia, the US and Russia) and the fourth largest oil exporter (behind Saudi Arabia, Russia and Norway). Its reserves are larger than those of the US and Russia but smaller than those of Saudi Arabia and of Iraq [15]. The oil and gas extracted from the Kurdistan region is used mainly for domestic consumption, while that from the north and south is exported.

Based on this discussion it is reasonable to state that Iran could be a candidate for changing EU's energy portfolio in a positive way. Iran with its large resources presents a new solution to the serious issue of security in the supply and sustainability of energy sources. This alternative seems promising as the current issues between Iran and the western powers are in the process of being solved through the recent agreement in Vienna on 14 July 2015. However, Iran's isolation and preventing international oil companies from investing in its oil and gas industry together with the country's reduction in the production of crude oil may hurt global energy security. In other words, the effect of the sanctions which have been imposed on Iran will influence the world energy market both in the short- and the long-term. The negative impact of these sanctions can be extended from a national to a global level in the future [35].

5. Competition Over Energy Resources by Superpowers

Russia has had the will to dominate Caucasus and Central Asian countries since the collapse of the Soviet Union. There are some crucial parameters for this policy. First, Caucasus has huge sources of energy and it is considered a bridge between Russia and Europe. In fact, Georgia connects Caucasus to Europe and plays an important role in energy transmission to Europe. The challenges in Caucasus regional security after the Russia-Georgia conflict in August 2008 and explained some risks associated with the operating transit energy corridor in southern Caucasus is studied [36]. The conflict was considered a part of Russia's willingness to rebuild its former control sphere. The North Atlantic Treaty Organization (NATO) is interested in extending its borders to western neighboring countries of Russia including Georgia and Ukraine. By this argument energy resources and expansion of NATO borders could be seen as the main reason for the Russia-Georgia conflict. 'Subduing Georgia would cut the west's vital energy connection (the Baku-Tbilisi-Çeyhan pipeline) to the Caspian Sea and to Central Asia' [37]. Afterwards, Russia will be able to dominate energy flows from Azerbaijan.

The proven oil reserves in various section of the Caspian Sea are different and are estimated at 298.4 billion barrels (Including Azerbaijan, Kazakhstan, Turkmenistan, Uzbekistan, Russia and Iran), corresponding to almost 17.5% of the total global proven reserves [16]. But if we consider the Azerbaijan, Kazakhstan, Turkmenistan and Uzbekistan segment it has 38.2 billion bbl of reserves. These figures make one doubt that Caucasus and Central Asia are major energy resources as compared to the Persian Gulf. The case is different when we consider Iran in our analysis because of its huge energy sources and geopolitical situation. The total reserves in Iran are estimated at 157.8 billion bbl which is more than Russia at 102.4 billion bbl and four times higher than Caucasus and Central Asian countries together. Therefore, the size of reserves in Caucasus and Central Asian countries together with Russia is not comparable with Iran. The high cost of oil exploration in Caucasus may not be economically beneficial as much as it is in Iran though. But the importance of the region in view of geopolitics, energy transmission and supply diversification cannot be insignificant.

The Baku-Tbilisi-Ceyhan and Baku-Supsa oil pipelines along with the Baku-Tbilisi-Erzurum gas pipeline have promoted the importance of Caucasus as an energy corridor between Asia and Europe. Russia's invasion of Georgia and unilateral recognition of Abkhazia and South Ossetia made a significant change in the Caucasus from a geopolitical point of view. Russia wanted to dominate Georgia in order to cut off Azerbaijan and Central Asian countries and consequently it was able to enhance its energy monopoly in Europe significantly [36]. Caucasus is like a magnetic field that Russia, Turkey and Iran are competing over with each other along with the United States and European Union. Russia was able to stop NATO's expansion into Caucasus by recognizing Abkhazia and South Ossetia [38].

Based on Henry Kissinger's thinking, capability and will are crucial variables in foreign affairs; although Russia was willing to dominate Caucasus countries but it did not have the capability [38]. Caucasus and Central Asian countries are strong enough to not be dominated by Russia. In this regard, Russia is being challenged by China and the western powers in this region. The geopolitical situation in Caucasus is important for Turkey because it is the connection point between Europe and Asia. Therefore, Turkey is going to develop its influence in this region. At the same time, Turkey competes with Russia and Iran for energy transmission to catch a share of Europe's energy market. Although Turkey does not have any energy resources but its role as an energy corridor to the west combined with a comprehensive investment program in refinery capacity are significant. The European Union relies on Russia, as it was the main source for 40% of its imported natural gas last year [16]. Therefore, European countries are looking for supply diversification through alternative routes such as Caucasus and Iran to improve their energy security.

6.   Isolation of Iran and Its Implications for Regional Security

Energy security as a dominant factor in international political stability is critical for major economics such as the United States, European Union, China and India. Any supply disruptions or fundamental changes may affect these economies through price volatility. Global energy markets have been affected by political instability and military conflicts in North Africa, the Middle East and Caucasus during the last decade. The Middle East is a formation of a multi-polar Regional Security Complex (RSC) that Iran plays as a regional power. Therefore, it is not possible to ignore it as an important factor for making and re-establishing a balance in the region. The Middle East and Caucasus faced imbalances after the Iran revolution in 1979 and RSC needs to be built in these regions through new policies. Sanctions imposed on Iran because of the country's nuclear activities targeted investments in oil, gas and petrochemicals, and exports of refined petroleum products. But in reality, Iran's political and economic isolation and the restrictions imposed on oil and gas field development can be considered a disruptive factor in global energy security.

Iran might have been an alternative energy source for EU when it faced a challenge with Russia regarding its intervention in Ukraine. However, EU lost this opportunity following Iran's isolation. After several rounds of nuclear talks between Iran and the western powers, UN sanctions were lifted on 16 January 2016. Consequently, Iran started increasing its oil production and exporting capacity in order to regain the market share that it had lost. Furthermore, Iran is going to introduce a new form of petroleum contracts in order to make them attractive for foreign investors. It is forecast that Iran will be a sustainable source of energy for the global market in the near future. Europe has the option to use the Iran-Turkey pipeline as an alternative for its natural gas transmission. But, Caucasus is another option available for exporting natural gas to Europe. In this regard, Caucasus has been the focus of attention of all parties including Europe, Russia, Turkey and Iran. Undoubtedly, Russia and Turkey are not interested in seeing enhanced regional importance for Iran.

Considering the tensions between Russia and Ukraine, it is clear that Russia is not viewed as a reliable sustainable energy source for Europe. But, Iran and Caucasus have enough capability individually or combined to be an alternative channel for energy transmission to Europe. In this regard, foreign investments and technology transfers are required for developing the oil and gas fields in Iran and for constructing the necessary energy transfer infrastructure. Also, Caucasian countries suffer from lack of energy infrastructure. Energy security is important both from the point of view of supply and demand. Therefore, it will be a win-win situation if major energy consuming EU countries undertake investments in energy supplier countries to build infrastructure, transmission and upstream and downstream facilities.

The military superiority of the US and the lack of a rival superpower to create a balance has led to replacement of global political space by a much more monolithic domination by the west. In the absence of an ideological challenger within or adjacent to the core, the Western powers in general and the US in particular, can impose much more demanding legal, social, financial, and political conditions [29]. Iran is considered as a regional power in Middle East and due to its critical role in views of energy security in the regional and global levels, its importance cannot be ignored. In order to convert an energy security complex to an energy security community, international cooperation among all parties is required. In our view, Western powers should take into account the fact that Iran can play a polar in regional stability and global energy security. Iran has a huge source of oil and gas and may be an effective alternative for Europe to achieve supply diversification in order to reduce its energy dependency to Russia. Therefore, international cooperation and performing JCPOA in case of Iran, may reduce the Iran-US conflict.

However, conservative policymakers in the United States supported by Israel and rival Arab countries in the Middle East have come up with additional impediments in rejoining Iran to the international community. It is clear that Iran has the will and the capability to play its role as a regional power in the Middle East. The country's open interventions in the security and stability affairs of Iraq, Syria and Yemen are evidence of its desire and capacity. Considering the geopolitics of Iran in the Middle East and its capabilities, it can be predicted that it will be a key factor in global energy security in the next decades as it has large energy reserves and a strategic position for energy transmission. Obviously, decisions made by Iranian policymakers and their interactions with western countries, particularly the United States, will be vital for this purpose.

7. Conclusion

There is a close relationship between diversifying the supply of energy sources and enhanced energy security. The Russia-Ukraine gas crisis not only illustrated to EU how strongly it is dependent on Russian gas, but also emphasized its need to diversify its energy supply sources. Iran and Caucasus are considered the main sources and routes for energy transmission to the global market, including to Europe. Security in Caucasus is affected by Russia's intervention in Georgia and the possibility of Armenia and Azerbaijan getting support from Russia, Turkey and Iran. Although NATO was interested in expanding its borders to include Caucasian countries but the Russia-Georgia conflict was an impediment in this. In the case of Iran, developing oil and gas fields, economic capacity and the country's international relations were significantly and negatively affected by the sanctions imposed against it by the United States and its European allies.

Conservative policymakers in the US supported by Israel and regional rivals like Saudi Arabia are going to keep Iran isolated after the lifting of UN sanctions. However, the geopolitical situation of Iran in the Middle East and its role as a regional power make achieving this target difficult. Russia is trying to rebuild the influence of the former Soviet Union but it does not have enough capacity to enforce this on states in Caucasus and Central Asia. The importance of energy transmission has converted Caucasus and Iran into a game of chess played by great powers like the US that have tried so as to bypass Iran and Russia by constructing the BTC pipeline. Considering Iran's geopolitics in the Middle East and its capabilities, it can be predicted that Iran will be a key factor in global energy security in the next decades due to its massive energy resources and strategic location for energy transmission. Caucasus too can be an alternative for natural gas transmission instead of Turkey. Therefore, the situation of Iran-Caucasus will be enhanced if this plan is realized in the future. International cooperation and performing JCPOA completely in case of Iran, may reduce parts of Iran-US conflict. Due to having a huge source of oil and gas, Iran has sufficient capability to serve as an effective alternative for Europe to achieve supply diversification in order to reduce its energy dependency on Russia. The combination of Iran resources and Caucasus as a corridor of energy supply could be a proper solution for enhancement of energy security in Europe. However, recent changes in US policy towards Iran is harmful for the EUs energy security.

Acknowledgments

The authors are grateful for comments and suggestion by an anonymous referee and editor of the journal on an earlier version of this manuscript.

Conflict of Interest

All authors declare no conflicts of interest in this paper.

References

1. Halder P, Prokop P, Chang CY, et al. (2012) International survey on bioenergy knowledge, perceptions, and attitudes among young citizens. *Bioenerg Res* 5: 247–261.

2. Hosseini SE, Andwari AM, Wahid MA, et al. (2013) A review on green energy potentials in Iran. *Renew Sust Energ Rev* 27: 533–545.

3. Nakata T, Silva D, Rodionov M (2011) Application of energy system models for designing a low-carbon society. *Prog Energ Combust Sci* 37: 462–502.

4. Tzanakis I, Hadfield M, Thomas B, et al. (2012) Future perspectives on sustainable tribology *Renew Sust Energ Rev* 16: 4126–4140.

5. Sigot M (2013) Energy security and the EU: between independency priorities, strategic vulnerabilities and sustainability needs. Master thesis, Program in Environmental Law, Stockholm University.

6. Bang HK, Ellinger AE, Hadjimarcou J, et al. (2000) Consumer concern, knowledge, belief, and attitude toward renewable energy: an application of the reasoned action theory. *Psychol Mark* 17: 449–468.

7. Cacciatore MA, Binder AR, Scheufele DA, et al. (2012) Public attitudes toward biofuels: effects of knowledge, political partisanship, and media use. *Polit Life Sci* 31: 36–51.

8. Schaeffer R, Szklo AS, Pereira de LAF, et al. (2012) Energy sector vulnerability to climate change: a review. *Energy* 38: 1–12.

9. Zyadin A, Puhakka A, Ahponen P, et al. (2012) School students knowledge, perceptions, and attitudes toward renewable energy in Jordan. *Renew Energ* 45: 78–85.

10. Barton B, Redgewell C, Ronnel A, et al. (2004) Energy security, managing risk, a dynamic legal and regulatory environment, Oxford University Press.

11. Halliday F (2005) The Middle East in international relations. *Contemp Rev* 10: 167–172.

12. Cordesman AH, Kleiber M (2006) CSIS: Asian conventional military balance. Available from: http://www.comw.org/cmp/fulltext/0606cordesman.pdf.

13. Bahgat G (2010) Iran's role in Europe's energy security: an assessment. *Iran Stud* 43: 333–347.

14. Koolaee E (2011) Iran's role in energy security at regional and global levels. *Iran Econ Rev* 15: 95–115.

15. Moradi M (2006) Caspian pipeline politics and Iran-EU relations. *Unisci Disc Pap* 2006: 173–184.

16. BP statistical review of world energy (2016). Available from: http://www. indiaenvironment portal.org.in/files/file/bp-statistical-review-of-world-energy-2016.pdf.

17. Kumar S, Kwon HT, Choi KH, et al. (2011) Current status and future projections of LNG demand and supplies: a global prospective. *Energ Policy* 39: 4097–4104.

18. BP statistical review of world energy (2015a). Available from: papers/bp-statistical- review-world-energy-june-2015-3/.

19. BP energy outlook 2035 (2015b). Available from: https://www.iamericas.org// presentations/BP_North_America.pdf.

20. Bilgin M (2009) Geopolitics of European natural gas demand: supplies from Russia, Caspian and the Middle East. *Energ Policy* 37: 4482–4492.

21. Barysch K (2007) Turkey's role in European energy security. Center for European Reform Essays. Available from: http://www.cer.org.uk (Accessed: 5 September 2016).

22. World Bank indicators (2016). Available from: http://data.worldbank.org/indicator/EG.IMP.CONS.ZS?

23. De Haas M, Tibold A, Cillessen V (2006) Geo-strategy in the South Caucasus: power play and energy security of states and organizations. Netherlands Institute of International Relations Clingendael.

24. Wittich A, Maas A (2009) Country case study: South Caucasus. Regional cooperation in the South Caucasus. Available from: https://www.adelphi.de/de/system /files/mediathek/bilder/us_038_-_regional_cooperation_in_the_south_caucasus_1.pdf.

25. Petersen A (2007) Integrating Azerbaijan, Georgia, and Turkey with the West: the case of the East-West transport corridor. Available from: https://csis-prod.s3.amazonaws.com/s3fs-public/legacy_files/files/media/csis/pubs/070910_petersen_commentary.pdf.

26. Chang CP, Berdiev AN, Lee CC (2013) Energy exports, globalization and economic growth: the case of South Caucasus. *Econ Model* 33: 333–346.

27. Bacik G (2006) Turkey and pipeline politics. *Turk Stud* 7: 293–306.

28. Argus News and Analysis, CPC expansion to be completed by year end (2017). Available from: https://direct.argusmedia.com/newsandanalysis/article/1397465.

29. Buzan B, Waever O (2003) Regions and powers: the structure of international security, Cambridge University Press.

30. İpek P (2006) The aftermath of Baku-Tbilisi-Ceyhan pipeline: challenges ahead for Turkey, Available from: http://sam.gov.tr/wp-content/uploads/2012/02/PinarIpek.pdf.

31. Cornell SE, Ismailzade F (2005) The Baku-Tbilisi-Ceyhan pipeline: implications for Azerbaijan, In: Starr SF, Cornell SE, *the Baku-Tbilisi-Ceyhan pipeline: oil window to the west*, Baku Press.

32. Auzer KA (2016) Institutional design and capacity to enhance effective governance of Iraqi-Kurdistan's oil and gas wealth. PhD thesis, Warwick Business School.

33. Koolaee E (2011) Iran's role in energy security at regional and global levels. *Iran Econ Rev* 15: 95–115.

34. Tofigh AA, Abedian M (2016) Analysis of energy status in Iran for designing sustainable energy roadmap. *Renew Sust Energ Rev* 57: 1296–1306.

35. Heshmati A, Abolhosseini S (2016) European energy security: challenges and green opportunities. *Work Pap*, 1–21.

36. Kakachia KK (2011) Challenges to the South Caucasus regional security aftermath of Russian-Georgian conflict: hegemonic stability or new partnership? *J Eurasian Stud* 2: 15–20.

37. Brzezinski Z (2009) Russia must re-focus with post-imperial eyes. Available from: http://tomweston.net/postimperial.pdf.

38. Matsuzato K (2010) Cultural geopolitics and the new border regions of Eurasia. *J Eurasian Stud* 1: 42–53.

6

Performance and emission reduction potential of micro-gasifier improved through better design

Kamil Dino Adem * and Demiss Alemu Ambie

School of Mechanical and Industrial Engineering, Addis Ababa Institute of Technology, Addis Ababa University, Addis Ababa, Ethiopia

* **Correspondence:** Email: kamil.dino@aau.edu.et;kdadem@gmail.com

Abstract: Biomass gasification is getting popular for household cooking application in most developing countries including Ethiopia. The preference for biomass gasification is due to the generation of less CO (Carbon Monoxide) and PM (Particulate Matter) in comparison with other biomass cookstoves. Our study showed the improvement in thermal efficiency and emission reduction potential of micro-gasifier. A prototype micro-gasifier was built and tested using the water boiling test protocol. The test results gave a thermal efficiency of 39.6% and a specific fuel consumption of 57 g of fuel/ liter of water. With regard to indoor air pollution, the maximum CO & PM registered were 12.5 ppm and 1.85 mg/m^3, respectively. Using clean development mechanism (CDM) methodology, the estimated emission reduction potential of the micro-gasifier is 1.30 tCO$_2$ per micro-gasifier per year. Generally, the micro-gasifier has better performance compared to the previous designs proposed by other researchers. Thus, disseminating our micro-gasifier at a larger scale in developing countries such as Ethiopia will be beneficial in reducing deforestation and emission that will be brought about by using open-fire stoves and thus, helps to obtain carbon credit.

Keywords: micro-gasifier; gasifier stove; biomass; performance evaluation; indoor air pollution; greenhouse gas

Abbreviations and Acronyms

AAiT	Addis Ababa Institute of Technology
AIT	Asian Institute of Technology
ASTM	American Society of Testing and Materials
CDM	Clean Development Mechanism
CO	Carbon Monoxide

ER	Emission Reduction
FCR	Fuel Consumption Rate
GHG	Greenhouse Gas
GTP	Growth and Transformation Plan
PM	Particulate Matter
SFC	Specific Fuel Consumption
SGR	Specific Gasification Rate
UNFCCC	United Nation Framework for Climate Change Convention
WBT	Water Boiling Test
WHO	World Health Organization
$B_{old,capita}$	Average baseline fuel wood consumption in tones per capita per year
$HC_{fuelwood,usage,y}$	Host country national fuel wood consumption in tones per year y
$B_{y,device}$	Average annual consumption of woody biomass per appliance in tones per year
$FW_{proportion}$	Proportion of household fuel wood consumed by micro-gasifier
$ER_{y,micro-gasifier}$	Emission reductions by project device during year y in tCO$_{2e}$ (tones of CO$_2$ equivalent)
$f_{NRB,y}$	Fraction of woody biomass saved by the project activity in year y that can be established as non-renewable biomass using default country specific fraction of non-renewable woody biomass
$NCV_{biomass}$	Net calorific value of the non-renewable woody biomass that is substituted
$EF_{projected\_fossilfuel}$	Emission factor for the substitution of non-renewable woody biomass by similar consumers
$N_{y,micro-gasifier}$	Number of project devices of type i=1 operating in year y

1. Introduction

In developing countries, 50% of the population depends on traditional use of biomass. More specifically, in Sub-Saharan Africa, around 753 million people (i.e. 80 percent of the population) use biomass as energy source. And 95 percent of the population living in Ethiopia uses biomass energy for household activities as well [1]. This will likely to continue in the future [2]. In developing countries, traditional biomass stoves are inefficient and are causing indoor air pollution. In order to curb this problem, gasifier stoves are becoming popular. The only improved biomass cookstove selected for dissemination in Ethiopia is *Tikikil* which has a thermal efficiency of 28% [3]. This cookstove saved on average 1.07 tons of CO$_2$ per stove where it is used in *kaficho* zone, Ethiopia [4].

Biomass gasification is considered as the future concept for cookstoves which are mostly forced type [5]. Currently, the number of patents obtained by companies for small-scale gasifiers are relatively low [6]. Reed and Larson [7] initiated the concept of gasification of wood gas stoves for cooking applications and the test result showed better performance in indoor air pollution and efficiency. Later on, a natural draft cross-flow gasifier tested at AIT (Asian Institute of Technology) also showed a thermal efficiency of 27% [8]. Other authors also reported a thermal efficiency of 26.5% and 35% with CO (3–6 ppm) and CO$_2$ (17–25 ppm) for gasifier stoves [9,10]. Dixit, et al. [11] has also reported a thermal efficiency of 37%.Mukunda, et al. [12] stated that forced draft gasifier stoves reached thermal efficiency of up to 50%. A number of efforts are underway to compare empirical result with computational models for gasifier stoves and generally, the results showed reasonable approximation between the two [13]. However, to increase the thermal efficiency and reduce the indoor air pollution remains to be researched. As a result, bringing a more efficient and less polluting gasifier stoves remain to be a challenge for researchers. In light of this, the objective of this study was to evaluate the performance of a micro-gasifier using woody biomass and compare its advantage in terms of indoor air pollution and generation of carbon credit.

2.   Materials and Method

2.1. Type of stove and materials for manufacturing

A gasifier stove as is shown in Figure 1 has been developed at AAiT (Addis Ababa Institute of Technology- Ethiopia). its primary air inlet consists of four holes with a diameter of 30 mm at 90^0 apart on the outer casing and 12 holes with a diameter of 4 mm at the bottom of the inner cylinder. The secondary air is used for combusting the gases produced from pyrolysis of the biomass. For doing this, there are 6 holes with a diameter of 4 mm at the top of the central pipe.

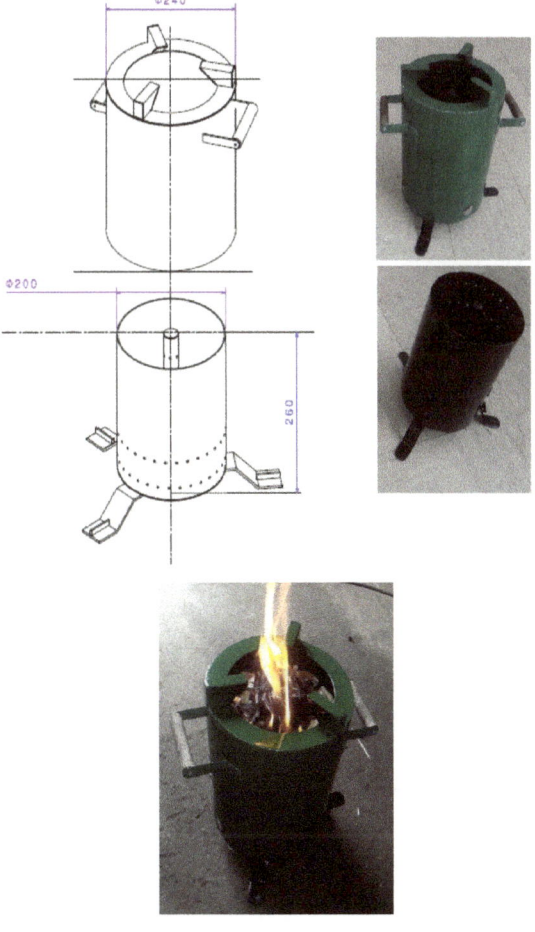

Figure 1. Micro-gasifier.

Most parts of the micro-gasifier such as: reactor, *outer casing, bottom cover and central pipe* were made of mild steel with a thickness of 1.5 mm. The handle is the only part which is made of wood.

2.2. Fuel used for testing

The type of fuel used for conducting the laboratory test was Eucalyptus tree which is commonly used in most part of Ethiopia. Proximate analysis was conducted to analyze the feed stock characteristics. The feed stock characteristics were, moisture content (ASTM D3173-73), volatile matter (ASTM D3175-73) and ash content (ASTM D3174-73). Fixed carbon (FC) was also determined using material balance. Physical and thermal properties of the biomass are shown in Table 1.

Table 1. Physical and thermal properties of eucalyptus tree as feed stock.

Characteristics	Biomass (Eucalyptus Tree)
Size (mm)	15–20
Length (mm)	30–50
Dry density (kg m^{-3})	480
Moisture content(% wb)	5.64
Volatile matter (% db)	80.81
Ash content (%db)	54
Fixed carbon (%db)	13.02
Calorific value (MJ kg^{-1})	18.64

2.3. Design of Micro-gasifier

To manufacture the micro-gasifier, it requires mild steel sheet metal with a thickness of 1.5 mm. The micro-gasifier was manufactured to be aesthetically appealing to the users. User satisfaction is one of the parameters used to evaluate stove acceptability at community level.

The design procedure outlined in [14] was used to come up with the major parameters required for drafting and manufacturing the gasifier stove. This includes *Energy Input, Reactor Diameter, and Height of reactor.*

Energy required for cooking (Energy demand): This refers to the amount of energy need for cooking of a kg of rice[1]. It can be computed by knowing the type of food to be cooked, the specific energy ($E_s = 331.8 \ kJ/kg$) and the time required ($T = 15 \ min$) for cooking.

$$Q_n = \frac{M_f \times E_s}{T} = 0.369 \text{ kW} \tag{1}$$

Energy input for cooking (Energy input): This is the amount of energy required for cooking and to be supplied for the heating application. The calorific value of the fuel is 18.64 MJ/kg. Assuming a stove efficiency of 17%, the fuel consumption rate of the fuel can be computed as follows:

---

[1] *Assuming most cooking foods in Ethiopia resemble rice.*

$$FCR = \frac{Q_n}{(HV_f)(\xi_g)} = 0.419 \text{ kg/hr} \tag{2}$$

Reactor Diameter: It is a major parameter of the micro-gasifier. It is the diameter of the reactor which is obtained by the ratio of FCR to SGR. Assuming FCR to be 2 kg/hr and SGR to be 120 kg/m^{2}hr, the reactor diameter is calculated by the following formula.

$$D = \sqrt{\frac{1.27 \, FCR}{SGR}} = 145 \text{ mm} \tag{3}$$

Hence, the diameter of the reactor is assumed to be 200 mm for this particular case in order to accommodate the variation of cooking habits throughout the country.

Reactor Height: This is the overall height of the reaction chamber. This dimension indicates the loading capacity of the reactor. It is calculated based on $\rho_a = 100 kg/m^3$.

$$H = \frac{SGR x T}{\rho_a} = 300 \text{ mm} \tag{4}$$

For this particular test, the height is reduced to 260 mm to compensate the increase in diameter.

2.4. Performance evaluation methods

The laboratory test performed to evaluate the stove performance was based on a revised version of the Water Boiling Test Protocol [15].

Thermal Efficiency: The amount of heat gained by the water inside the pot and evaporated compared to the energy of the fuel used for heating. The value for the efficiency can be computed based on the following equation:

$$\eta_{th} = \frac{m_w * C_p * (T_f - T_i) + m_{ev} * h_{fg.w}}{m_f * LHV_f} \tag{5}$$

Burning rate: The rate of fuel consumption during which the water gets boiled. It is calculated by dividing the equivalent dry fuel consumed by the time of the test.

$$r_{cb} = \frac{f_{cd}}{\Delta t_c} \tag{6}$$

Specific fuel consumption: The amount of fuel consumed per unit mass of boiled water. It is calculated based on the following formula

$$SFC = \frac{f_{cd}}{m_w} \tag{7}$$

Fire power: This is the fuel energy consumed to boil the water divided by the time to boil. It tells the average power output of the stove (in Watts) during the high-power test.

2.5. Instrumentation for measurement

Figure 2 shows the test set-up for conducting the water boiling test and the corresponding equipment required for conducting the test. The instruments used for measuring the temperature were

K-Type thermocouple and an infrared thermometer for measuring flame temperature. In addition, sensitive balance was used for measuring the weight of water and fuel. A standard pot commonly used in Ethiopia was used to evaluate the performance of the micro-gasifier which is 7 liter capacity stainless steel pot with 5 liter of water.

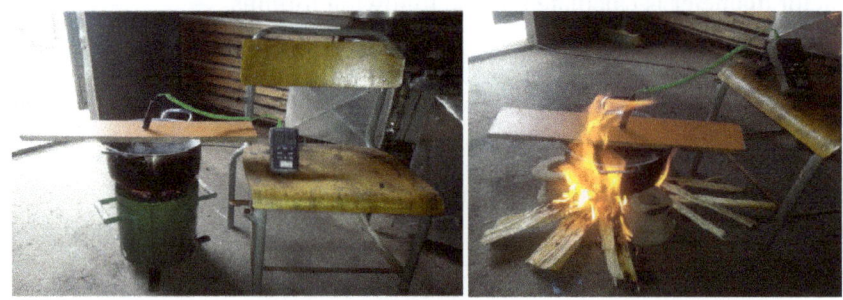

Figure 2. Test set-up of micro-gasifier and open-fire stove.

Figure 3 below shows the feedstock, sensitive balance and thermometer which were used to evaluate the performance of the micro-gasifier. The standard water boiling test protocol was used to obtain the thermal efficiency of the micro-gasifier.

Figure 3. Wood and sensitive balance used for the test.

Carbon monoxide and particulate matter concentrations as well as room temperataure were logged every minute during sampling period of the laboratory test. The test room is selected in such a way that it simulates a typical rural household kitchen (2 m × 2 m area and 2.5 m-height). The instruments used for measurement were CO data logger with USB interface (EL-USB-CO with accuracy of ±6 ppm) for carbon monoxide; UCB Particulate Monitor (University of California,Berkeley) for particulate matter ($PM_{2.5}$); and thermocouple data logger with USB interface (EL-USB-TC-LCD with accuracy of ±1 °C) for temperature. The procedure used to set the above cited instruments were based on the procedure outlined by the University of California-Berkeley. The standard procedure for installing the indoor air pollution (IAP) instruments is approximately 100 cm from the edge of the combustion zone and a height of 145 cm from the floor. Here, it is also recommended to place the instruments at least 150 cm away (horizontally) from openable doors and windows where possible [16].

3. Results and Discussion

3.1. Thermal efficiency and specific fuel consumption

The final output of the micro-gasifier designed and manufactured at Addis Ababa Institute of Technology (AAiT) is shown in Fig.4 with the final paints to make it appealing to the end users.

Figure 4. Designed and manufactured micro-gasifier.

Thermal efficiency is calculated based on output over input heat energy. The output is related to the amount of heat supplied to the water in the pot and the input is the energy supplied to the water from the wood burned in the micro-gasifier. The average thermal efficiency obtained is 39.6 ± percent. The other performance indicator is the specific fuel consumption, which shows 70 percent improvement compared to the open fire stove (Table 2).

Table 2. Performance Parameters for Micro-gasifier.

No.	Test parameters	High Power Test (Cold Start)					
		Micro-gasifier					Three-stone
		Test-1	Test-2	Test-3	Avg.	Standard Error	Test-1
1.	Time to boil Pot #1 (min)	30	32	33	31.7	0.87	24
2.	Temp-corrected time to boil Pot #1 (min)	34	34	34	34.0	0.23	27
3.	Burning rate (g/min)	9	9	7	8.0	0.81	35
4.	Thermal efficiency (%)	37	35	47	39.6	3.76	14
5.	Specific fuel consumption (g/liter)	58	65	49	57.0	4.62	187
6.	Temp-corrected specific fuel consumption (g/liter)	66	68	50	61.0	5.83	213
7.	Temp-corrected specific energy cons. (kJ/liter)	1247	1285	940	1158.0	109.4	4013
8.	Firepower (watts)	2822	2954	2158	2644.7	246.2	10913

Bhattacharya and Leon [8] and Belonio [14] obtained a thermal efficiency of a gasifier stove as 27% and 25.9%, respectively. Dixit, et al. [11] also reported thermal efficiency of a gasifier stove as 37%. Panwar and Rathore [9] and Panwar [10] obtained 26.5% and 35%, respectively. The thermal efficiency and specific fuel consumption of our micro-gasifier showed improvement compared to the existing gasifier stoves in the market (Table 2).

As it is usually expected, the time required for boiling 5 liter of water in Addis Ababa using micro-gasifier is 70 percent higher than using the open-fire stove. The Firepower of the micro-gasifier is around 2.60 kW. In this study, the feedstock was limited to woody biomass in which case the inclusion of other feedstock may bring a different value in terms of thermal efficiency and indoor air pollution.

3.2. Boiling point of water in Addis Ababa-Ethiopia – AAiT Workshop

In order to ensure the precision of computing the performance of the micro-gasifier, the boiling point of water was experimentally determined rather than calculating using the altitude. The mean boiling point of water in Addis Ababa was 91 °C where the altitude is 2355 m (Figure 5). As shown in Figure 5, three tests were conducted to determine the average boiling temperature of water in a specific location.

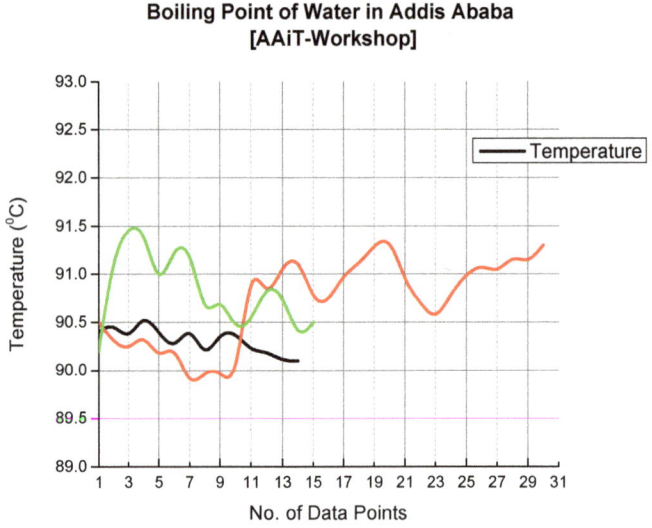

Figure 5. Boiling point of water in Addis Ababa.

3.3. Indoor air pollution

The major two emissions considered for this test in relation to indoor pollution were CO and PM. Figure 5 shows the variation of CO during the test period. The maximum CO exhibited during this test was 12.5 ppm which is below the WHO standard set as 30.6 ppm for 1 hour exposure [17]. As shown in Figure 5, the CO measurements exhibited the maximum value during the 40–50 minutes cooking time.

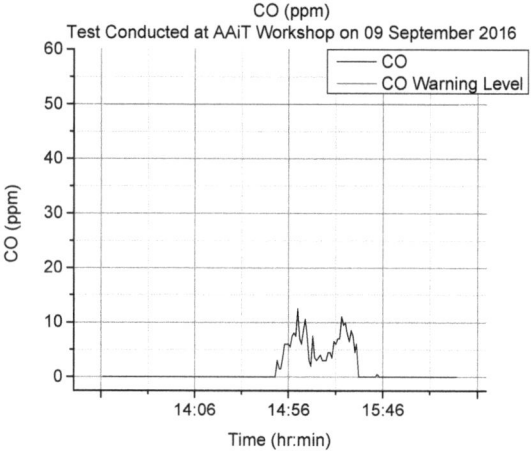

Figure 6. CO concentrations during testing.

Similarly, the amount of PM generated during the test period is plotted to see whether it is in the acceptable range or not. The maximum amount of PM recorded was 1.85 mg/m^3 during the testing period (Figure 7). WHO guideline indicates that the annual mean & 24-hour mean for PM$_{2.5}$ is 10 mg/m^3 and 25 mg/m^3, respectively [18]. Thus, the experimental result for PM$_{2.5}$ is well below the recommended value. The reduction in PM of micro-gasifier showed 10–12 folds which is a huge advantage in terms of indoor air pollution, making the stove highly acceptable by the users.

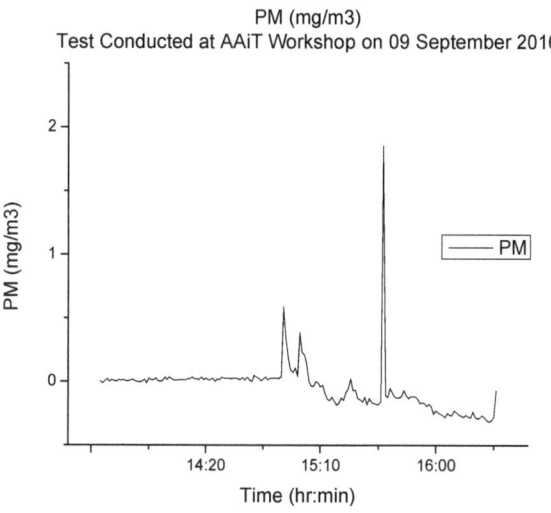

Figure 7. PM concentrations during testing.

Indoor air pollution is responsible for the death of many people in developing countries during cooking. This can be substantially reduced by improving and introducing micro-gasifier. By using a better design of micro gasifier as shown in our case, it was possible to reduce the two emissions that are good indicator of indoor air pollution, CO and PM-where they were significantly reduced below WHO standard [16].

The room used for conducting the test was a simulated typical Ethiopian rural household in which case the door was completely open to the outside. There were also some openings in the upper part of the room. Figure 8 shows the variation of the room temperature with time during the operation of the stove.

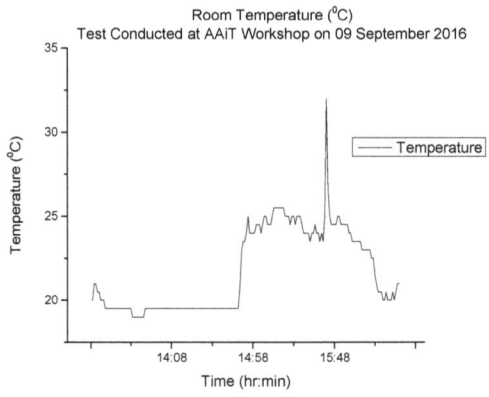

Figure 8. Room temperature vs time.

3.4. Time vs temperature variation for various tests

The variation of temperature as heating was continued until the water reached its boiling point was shown in Figure 9. In the four testes conducted, the temperature exhibited a similar trend

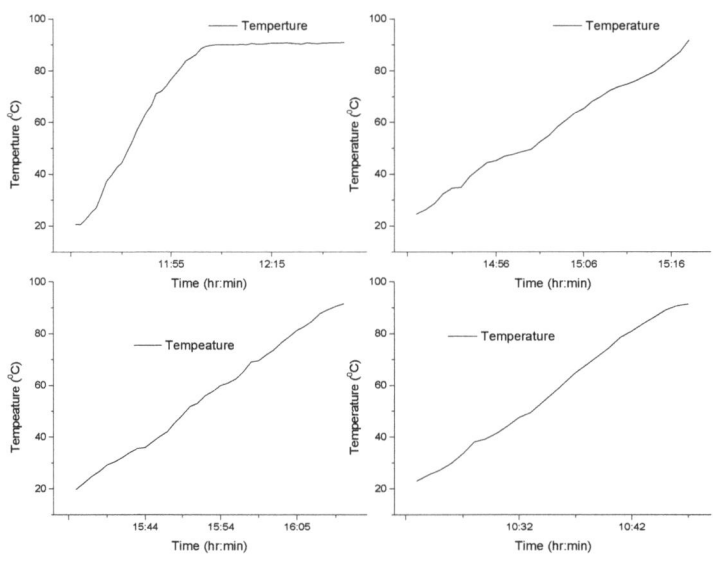

Figure 9. Temperature of water during heating process.

showing the consistency of the boiling of the water at the given altitude and pressure. Similarly, the result indicates the precision level of our test result.

3.5. Conservative estimate for GHG reduction potential

As we have mentioned above, the majority of the population in Ethiopia uses biomass energy for cooking. In order to estimate the GHG saving from introducing micro-gasifiers, the reference dissemination potential of households could be considered based on GTP 2 of Ethiopian Government Plan. In this plan, 11.25 million improved cookstoves are expected to be distributed in the coming five years. The discussion under section 3.5. focuses on estimation of the amount of CO_2 reduced per unit of micro-gasifier, where the total amount of CO_2 reduction can be computed based on baseline survey for a specific location.

The approach followed to calculate the amount of emission reduced per unit of micro-gasifier was based on the clean development mechanism (CDM) methodology outlined by UNFCCC[19].

$$B_{old,capita} = \frac{HC_{fuelwood,usage,y}}{HC_{population,y}} \tag{8}$$

$$HC_{fuelwood,usage,y} = Fuel\ wood\ consumption\ in\ cubic\ meters\ x\ Wood\ density \tag{10}$$

$$Fuel\ wood\ consumption\ in\ cubic\ meters\ = 80{,}185{,}000\ m^3\ [20]$$

$$Wood\ density\ = 0.725\ t/m^3\ [21]$$

$$HC_{fuelwood,usage,y} = 80{,}185{,}000\ m^3\ x\ 0.725\ t/m^3 = 58{,}134{,}125\ t/year$$

$$HC_{population,y} = 84{,}320{,}987\ [22]$$

$$B_{old,capita} = 0.689\ t/year$$

$$N_{residents,household} = 6$$

$$B_{y,device} = B_{old,capita}\ x\ N_{residents,household}\ x\ FW_{proportion} \tag{11}$$

$$FW_{proportion} = 41.50\%\ for\ cooking\ application$$

$$B_{y,device} = 1.72\ t/device/year$$

$\eta_{old} = 10\%$ [The default value of 10% was applied as the systems to be replaced are three stone fires].

$$\eta_{new,micro-gasifier} = 39.6\%$$

$$B_{old,micro-gasifier} = B_{y,deivce,micro-gasifier}\ x\ L_y \tag{12}$$

The default net to gross adjustment factor of 0.95 has been applied to account for leakages:

$$B_{old,micro-gasifier} = 1.63\ t/device/year$$

$$B_{y,savings,micro-gasifier} = B_{old,micro-gasifier}\ x\ \left(1 - \frac{\eta_{old}}{\eta_{new,micro-gasifier}}\right) \tag{13}$$

$$B_{y,savings,micro-gasifier} = 1.21 \frac{t}{device}/year$$

The emission reductions created by the micro-gasifier are calculated as follows:

$$ER_{y,micro-gasifier} = B_{y,savings,micro-gafier} \; x \; f_{NRB,y} \; x \; NCV_{biomass} \; x \; EF_{projected\,fossil\,fuel}$$

$$x \; N_{y,micro-gasifier} \tag{14}$$

$f_{NRB,y} = 88\%$ [23]

$NCV_{biomass} = 0.015$ TJ/tone (IPCC default for wood fuel on wet basis)

$EF_{projected\_fossilfuel} = 81.6$ tCO2/TJ

$$\boldsymbol{ER_{y,micro-gasifier} = 1.30 \; tCO_2/device/year}$$

The emission reduction potential of the micro-gasifier, in terms of carbon credit saving which was calculated as 1.30 tCO$_2$/device/year is better than other biomass cookstove promoted in the country by various institutions. For instance, World Vision Ethiopia claimed 1.14 tCO$_2$/device/year [24].

Our result showed better emission reduction which implies that the micro-gasifier tested in our study has a bigger contribution towards CO$_2$ emission to the environment, thereby reducing global warming and climate change.

3.6. Manufacturing cost of micro-gasifier

The micro-gasifier was made of mild steel sheet metal with 1.5mm thickness and assumed that it will be used at least for five years without maintenance. The unit production cost of the micro-gasifier is 10.85 USD (Table 3). This cost could be partially covered from the carbon saving of the micro-gasifier which is 1.3 tCO$_2$/device/year. For wider dissemination of the micro-gasifier, awareness creation about the benefit and its implication for alleviating the health problems need to be carried out through joint efforts of governmental and non-governmental and organizations.

Table 3. Manufacturing cost of micro-gasifier.

No.	Description	Total cost (USD)
1	Mild steel sheet metal 1.85m x 0.50m x 1.5mm	4.50
2	Pipe 3/4" and 260mm	0.35
3	Mild steel sheet metal 0.131m x 0.045m x 4.0mm	0.35
4	Rectangular Hollow Section(RHS) 20mm x 30mm x 1.5mm and length 180mm	0.35
5	Wood(for handle) Dia. 20mm and 110mm	0.05
6	Machining cost (cutting, grinding, drilling, etc...)	3.75
7	Labour cost	1.50
	Total	**10.85**

4. Conclusion

Based on our study, a gasifier stove is worth promoting rather than biomass cookstoves and open-fire stoves. This is due to its thermal efficiency and less indoor air pollution and emission reduction. Disseminating this micro-gasifier will result in a reduced fuel wood use in the rural parts of Ethiopia where local people widely use open-fire stove for household cooking. It also reduces indoor air pollution which is the cause of death incidents; and reduction of CO_2 emission which has tremendous implications for the environment in terms of global warming and climate change. The cost of micro-gasifier is relatively small and even that could be subsidized by having a partial carbon credit by selling the emission reduced. Such experience is already available within the country with some non-governmental organization like World Vision Ethiopia. Hence, this micro-gasifier is a better choice for the users.

Acknowledgments

This research was funded by Addis Ababa University (AAU) and Ethiopian Climate Innovation Center (ECIC). We would like to thank the Ministry of Water, Irrigation and Electricity and Ethio Resource Group (ERG) PLC for lending us the instruments which we used for various measurements during the test period.

Conflict of Interest

All authors declare no conflict of interest in this paper.

References

1. IEA (2015) World Energy Outlook. Population relying on traditonal use of biomass for cooking in 2013.
2. Beyene AD, Koch SF (2013) Clean fuel-saving technology adoption in urban Ethiopia. *Energy Economics* 36: 605-613.
3. GIZ-ECO, D.G.f.I.Z.-E.C.O. (2009) Water Boiling Test Results of Various Types of Household Wood Stoves for Non-Injera Cooking.
4. Dresen E, DeVries B, Herold M, et al. (2014) Fuelwood Savings and Carbon Emission Reductions by the Use of Improved Cooking Stoves in an Afromontane Forest, Ethiopia. *Land* 3: 1137.
5. Sutar KB, Kohli S, Ravi MR, et al. (2015) Biomass cookstoves: A review of technical aspects. *Renew Sust Energ Rev* 41: 1128-1166.
6. Kirkels AF, Verbong GPJ (2011) Biomass gasification: Still promising? A 30-year global overview. *Renew Sust Energ Rev* 15: 471-481.
7. Reed TB, Larson R (1996) A wood-gas stove for developing countries. *Energ Sust Develop* 3: 34-37.
8. Bhattacharya SC, Leon MA (2003) A Biomass-Fired Gasifier Stove (IGS-2) For Institutional Cooking. Available from: http://www.faculty.ait.asia/kumar/rets/Publications/Glow_Indonesia.pdf. [accessed 29.09.2016]

9. Panwar NL, Rathore NS (2008) Design and performance evaluation of a 5 kW producer gas stove. *Biomass Bioenerg* 32: 1349-1352.

10. Panwar NL (2009) Design and performance evaluation of energy efficient biomass gasifier based cookstove on multi fuels. *Mitigation and Adaptation Strategies for Global Change* 14: 627-633.

11. Dixit CSB, Paul PJ, Mukunda HS (2006) Part I: Experimental studies on a pulverised fuel stove. *Biomass Bioenerg* 30: 673-683.

12. Mukunda H, Dasappa S, Paul PJ, et al. (2010) Gasifier stoves - science, technology and field outreach. *Curr Sci* 98: 627-638.

13. Varunkumar S, Rajan NKS, Mukunda HS (2012) Experimental and computational studies on a gasifier based stove. *Energ Convers Manage* 53: 135-141.

14. Belonio AT (2005) *Rice Husk Gas Stove Handbook.*

15. WBT, v., *Water boiling test v.4.2.3.* Cookstove Emissions and Efficiency in a Controlled Laboratory Setting, 2013.

16. University of California, B. *Installing Indoor Air Pollution Instruments in a Home, Version 5.1.* 2005 Available from: http://berkeleyair.com/wp-content/publications/guidelines-for-instrument-placement.pdf. [accessed 10.08.2016]

17. WHO, R.O.f.E. (2010) WHO guidelines for indoor air quality: selected pollutants. Available from: http://www.euro.who.int/__data/assets/pdf_file/0009/128169/e94535.pdf. [accessed 03.09.2016]

18. WHO (2014) WHO indoor air quality guidelines: household fuel combustion. Available from: www.who.int/indoorair/guidelines/hhfc/HHFC_guidelines.pdf. [accessed 03 September 2016]

19. UNFCCC (2012) CDM, Small-Scale Methodology: Energy Efficiency Measures in Thermal Applications (AMS-II.G). Available from: https://cdm.unfccc.int/methodologies/DB/DCO8WRR QVTGLH1GHQBCL035F5M13R8. [accessed 19.09.2016]

20. UN, U.N.S.D. (2013) *Energy Statistics Database.* Available from: http://data.un.org/Data.aspx?d=EDATA&f=cmID%3aFW%3btrID%3a06. [accessed 28.09.2016]

21. FAO (2004) *Unified Bioenergy Terminology.* Available from: http://www.fao.org/docrep/009/j8227e/j8227e11.htm#P1131_70563. [accessed 19.09.2016]

22. Ethiopia, C.S.A.-. (2012) *Population Statistics Abstract.* Available from: http://www.csa.gov.et/images/documents/pdf_files/nationalstatisticsabstract/2011/2011%20popul ation.pdf. [accessed 28.09.2016]

23. UNFCCC. (2012) *Default values of fraction of non-renewable biomass.* Available from: http://cdm.unfccc.int/DNA/fNRB/index.html. [accessed 28.09.2016]

24. World Vison Ethiopia, W. (2013) *Energy Efficient Stoves Program, CDM Program of Activities Design Document.* Available from: https://cdm.unfccc.int/ProgrammeOfActivities/poa_db/5TE6 HLP1Z4KOABSDI873YQCFGXW2RM/view. [accessed 24.09.2016]

7

Assessment of offshore wind power potential in the Aegean and Ionian Seas based on high-resolution hindcast model results

Takvor Soukissian [1,*], Anastasios Papadopoulos [2], Panagiotis Skrimizeas [3], Flora Karathanasi [1,4], Panagiotis Axaopoulos [5], Evripides Avgoustoglou [3], Hara Kyriakidou [1], Christos Tsalis [6], Antigoni Voudouri [3], Flora Gofa [3], and Petros Katsafados [7]

[1] Institute of Oceanography, Hellenic Centre for Marine Research, Anavyssos, Greece
[2] Institute of Marine Biological Resources and Inland Waters, Hellenic Centre for Marine Research, Anavyssos, Greece
[3] Hellenic National Meteorological Service, Hellinikon, Athens, Greece
[4] Department of Naval Architecture and Marine Engineering, National Technical University of Athens, Zografos, Athens, Greece
[5] Department of Geology, Centre for Arctic Gas Hydrate, Environment and Climate, UiT, The Arctic University of Norway, Norway
[6] Atmospheric Modeling and Weather Forecasting Group, Division of Applied Physics, School of Physics, University of Athens, Athens, Greece
[7] Department of Geography, Harokopio University, Athens, Greece

* **Correspondence:** Email: tsouki@hcmr.gr

Abstract: In this study long-term wind data obtained from high-resolution hindcast simulations is used to analytically assess offshore wind power potential in the Aegean and Ionian Seas and provide wind climate and wind power potential characteristics at selected locations, where offshore wind farms are at the concept/planning phase. After ensuring the good model performance through detailed validation against buoy measurements, offshore wind speed and wind direction at 10 m above sea level are statistically analyzed on the annual and seasonal time scale. The spatial distribution of the mean wind speed and wind direction are provided in the appropriate time scales, along with the mean annual and the inter-annual variability; these statistical quantities are useful in the offshore wind energy sector as regards the preliminary identification of favorable sites for exploitation of offshore wind energy. Moreover, the offshore wind power potential and its variability are also estimated at 80 m height above sea level. The obtained results reveal that there are specific areas in the central and the eastern Aegean Sea that combine intense annual winds with low variability; the annual offshore wind power potential in these areas reach values close to 900 W/m^2, suggesting that a detailed assessment of offshore wind energy would be worth noticing and could lead in attractive investments. Furthermore, as a rough estimate of the availability factor, the equiprobable contours of the event [4 m/s ≤ wind speed ≤ 25 m/s] are also estimated and presented. The selected lower and upper bounds of wind speed correspond to typical cut-in and cut-out wind speed thresholds, respectively, for commercial offshore wind turbines. Finally, for seven offshore wind farms that are at the concept/planning phase the main wind climate and wind power density characteristics are also provided.

Keywords: wind speed-wind direction; offshore wind potential; variability; long-term hindcast simulations; Aegean and Ionian Seas

1.   Introduction

Offshore wind sector has become a primary energy policy among several European countries for decreasing carbon dioxide (CO_2) emissions and attenuating climate change impacts, as was set by the EU key targets for the year 2020 and 2030. As a result, the corresponding share of the annual EU wind energy installations was increased from 13% in 2014 to 24% the next year [1]. In line with this target, offshore wind energy has gained ground against onshore wind installations during the past few years.

Offshore and onshore wind conditions have great distinctions as regards development and installation of wind farms; for instance, offshore winds are higher in magnitude and more stable (spatially and temporally) than onshore winds producing considerably higher power over time while there are available wide-open offshore areas in contrast with the satiated onshore spaces for large-scale wind projects adjacent to energy demanding centers. On the other hand, offshore wind projects share a higher total cost for their materialization than onshore projects and raise greater risks and uncertainties due to the immaturity of technology and the lack of knowledge as regards the impacts of offshore wind farms (OWFs) on the socio-economic and environmental features of a candidate area; see also [2–5]. Currently, the most rapidly growing sea basin in terms of offshore wind installations is the North Sea with 69.4% of offshore wind capacity by 2015 [6], mainly due to the favorable combination of high wind potential and shallow water depths at great distances from the shore. In the Mediterranean Sea, numerous offshore wind projects have been planned, however immediate implementation of OWFs is not foreseen yet.

One of the most important and fundamental problems in the design and development of an offshore wind project is the preliminary identification of appropriate areas that meet specific criteria. The selection of appropriate sites is neither a simple nor a straightforward process; on the contrary, the determination of the suitability of an OWF site is a multilateral and complex procedure based on various considerations that are not always aligned. Site selection is, in general, a three-fold procedure built on the following distinct frameworks: i) the technical/engineering framework, which is relevant with the assessment of technical and engineering aspects (such as wind resource availability, water depth and bottom suitability, distance from shore, etc.), ii) the environmental framework concerning sensitive marine habitats and protected areas, effects and impacts on seabirds, fish and marine mammals, etc. and iii) the socio-economic framework dealing with the effects on human life, other coastal and marine activities/uses (e.g. tourism, fisheries, shipping lanes, marine cultural heritage, military exercise areas), financial issues, European and national legislation, marine spatial planning, etc. An application integrating some of the above-mentioned factors into a single framework is given in [7].

Apart from wind availability, distance from shore and water depth are currently the most important technical and financial constraints in the selection of sites for OWF development. As distance from shore and bottom depth increase, the costs of underwater electrical grid connections, installation and maintenance also increase. On the other hand, a short distance from shore causes visual disturbances in the neighboring coastal areas. The importance of visual disturbance is crucial for areas of high touristic development like the majority of the Mediterranean coasts [8,9]. For example, due to the strong reaction of the coastal communities against fixed installations, France has recently considered, at a regional level, to develop OWFs based on floating wind turbines that can be deployed in deeper water depths and off the Mediterranean coasts; see [10], as well as the published planning document by the Préfet Maritime de la Méditerranée and the Préfet de la region-Provence-Alpes-Côte d'Azur [11]. Moreover, three demonstration floating wind farms in the Mediterranean have been awarded from the French Ministry of the Environment, Energy and Marine Affairs, see e.g. [12]. It is anticipated that, if this strategy is strongly supported and successfully adopted in France, it will probably have implications on national policies and environmental, socio-economic, technical and legislative considerations of other EU Mediterranean countries including Greece. Moreover, on the presumption that the technology readiness level of the, still immature, floating wind turbines will reach the final stage and will be headed towards commercialization, and the corresponding costs will be significantly reduced as well (for instance, the cost of the floating foundations are four to six times higher than the fixed ones [13]), the installation of bigger floating turbines far from the coastline may provide a viable solution to the visual disturbance problem and weaken the Not-In-My-Backyard (NIMBY) attitude. Conversely, in Westerberg et al. [14], it is suggested that the attitude of the tourist community may be influenced as regards OWFs that are visible from shore if the proper information on offshore wind energy is provided (e.g. effectiveness of renewable energies, actual cost of offshore wind energy compared to conventional fuels in terms of climate change, etc.). A feasibility study including an economic evaluation, social benefits and a cost-benefit analysis of the floating wind technology in the Mediterranean Sea is presented in [15].

Wind resource assessment studies require reliable long-term measurements of wind speed and direction. However, *in situ* measurements are inadequate for such extended areas like the Mediterranean Sea. Thus, the only solution to overcome the hindrance of spatial coverage is the use of high-resolution gridded wind data sets in order to estimate accurately the parameters of interest even at nearshore areas.

Although there are several recent research studies mapping offshore wind resource over the Mediterranean Sea based on various wind data sources (e.g. [16–20]), in order to identify hot-spots for OWF development, the Greek Seas have received little attention despite their favorable wind climatology with respect to the development of such projects, especially in deep water depths. For instance, in Kotroni et al. [18], the wind climatology of Greece for different height levels is presented based on simulations of the high-resolution MM5 model with 2 km × 2 km grid resolution since 2000. After the verification of the model results with onshore meteorological stations, annually and monthly statistics of wind speed are provided over the mainland and maritime areas of Greece, as well as annual wind potential density at 50 m height. The SKIRON atmospheric model, a high-resolution atmospheric modeling system, was adopted by Emmanouil et al. [19] for the simulation of wind speed and direction between 2001 and 2010 with 0.05° × 0.05° horizontal resolution. The obtained result regarding the spatial pattern of wind power potential resembles the corresponding one from Kotroni et al. [18] reaching annual values around 600 W/m^2 (at 10 m height) in the eastern Aegean Sea. However, none of the above studies includes the spatial distribution of basic statistics of wind direction nor variability measures that are very useful in wind energy industry. The significance of assessing wind direction as well has been highlighted, among others, in [20,21,22] and is attributed to the minimization of wake effects in varying wind directions during the wind farm micrositing design. On the other hand, variability measures, such as mean annual and inter-annual variability, are essential parameters for the wind energy economics as was stressed by Watson [23] and references cited therein.

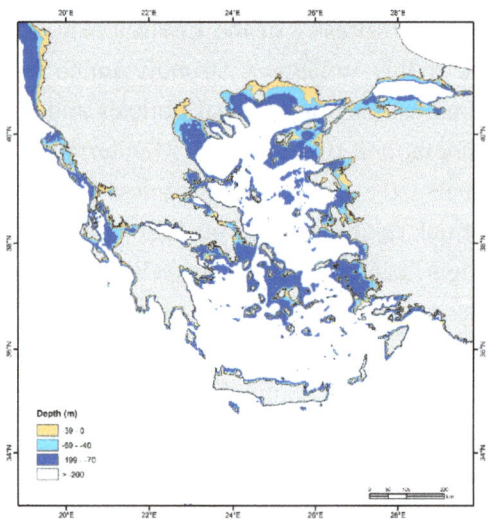

Figure 1. Bathymetry of the Aegean and Ionian Seas for specific water depth ranges (source of bathymetric data: EMODnet Bathymetry Portal [24]).

Nevertheless, OWF development in Greece is still at an infant stage. A rather important hindrance refers to the geomorphological features of the Greek Seas and, in particular, to the unfavorability of the bathymetric conditions. Specifically, the Greek continental shelf is very narrow; consequently, the effective water depths for OWF installations are located very close to the shore, increasing thus NIMBY attitudes. In Figure 1, a bathymetric map of the examined area is presented including specific depth ranges: i) 0–40 m ("shallow waters"), ii) 40–70 m ("intermediate waters"), and iii) 70–200 m ("deep waters"). Clearly, shallow and intermediate waters refer to monopile, gravity-based, tripod, jacket, and tri-pile supporting structure of offshore wind turbines. Deep waters (70–200 m) that have the highest spatial coverage over the examined sea areas, refer to floating wind turbines technologies, including tension leg platforms. Extended shallow water (0–40 m) areas are located in Northern Greece and at the north-eastern coasts of Limnos Isl. From this figure it is evident that bottom depth and, consequently, distance from shore are probably the most important technical parameters that should be considered in the development of OWFs in Greece.

This unfavorable situation (i.e. lack of Greek offshore wind projects) is also attributed to the financial crisis that, among others, prevented investments and the, until recently, unstable feed-in-tariff (FiT) scheme that caused additional doubt about the offshore industry and uncertainty to the potential investors; see, for example, [25]. However, on August 9, 2016 Law 4414/2016 was published in the Government Gazette [26], mainly containing a new support scheme for electricity from renewable energy sources (RES). This scheme aims to restructure the existing support scheme in order to promote better integration and participation of power generation from RES in the national electricity market, being also compatible with the European Commission's Guidelines on State aid for environmental protection and energy 2014–2020; see [27]. This new law introduces feed-in premiums (FiP), competitive tenders and virtual net metering. It is anticipated that Law 4414/2016 will rearrange the national energy sector by making relevant investments much more attractive; thus rendering OWF development in the Aegean and Ionian Seas more prosperous in the next few years.

The main objective of the present work is to provide the spatial distribution of offshore wind speed, wind direction and wind power density on an annual and seasonal scale over the Aegean and Ionian Seas, i.e. for two major sea areas where applications for production licenses have been submitted to the Greek Regulatory Authority for Energy (RAE). For this purpose, hindcast wind data, including wind speed and wind direction, with high spatial ($0.1° \times 0.1°$) and temporal (3 h) resolution covering the period from 1995 to 2009 has been validated and analyzed. Moreover, variability measures of wind speed and direction are also provided for the examined areas leading to a more complete description of offshore wind climatology. Estimates of the equiprobable contours of the event [4 m/s $\leq$ wind speed $\leq$ 25 m/s] that resembles the wind turbine availability factor are also quantified and presented in an annual basis. The particular wind speed threshold values correspond to typical cut-in and cut-out wind speeds for most commercial offshore wind turbines. Finally, for selected potential offshore wind energy projects that are at the concept/planning phase the main wind climate wind power density characteristics are also provided.

The structure of this work is the following: Section 2 deals with the main methodological aspects followed in this work, including the validation procedure of the model results (regarding both wind speed and wind direction) using collocated wind data obtained from oceanographic buoys. Section 3 is focused on the description of the numerical model and the presentation of the results from the statistical analysis and the validation procedure. In Section 4 the detailed numerical results referring to the spatial distribution of wind speed and direction, as well as wind power density, are presented. Specifically, the main statistical parameters of the above-mentioned quantities along with some important variability characteristics and wind speed persistence analysis are provided. Moreover, the particular characteristics of some selected locations of the Aegean and Ionian Seas, where OWF development is at the concept/early planning phase are also presented. In the last section, the key findings of the presented analysis are summarized.

2. Methodology

In this work, the wind climate analysis and the assessment of offshore wind power potential are based on a high-resolution gridded wind dataset obtained from a numerical weather prediction (NWP) model.

Model outputs are subject to various uncertainties, mainly attributed to: i) the boundary and initial conditions that are not accurate enough due to measurement and assimilation errors; ii) the exclusion of specific interactions or less important variables for simplification reasons, and; iii) the chaotic nature of the atmosphere that renders the modelling of this system a very demanding and exacting procedure; for this reason, the validation of the model results is necessary. In this respect, measured wind speed and wind direction data, obtained from offshore oceanographic buoys, were used.

As regards wind speed, the statistical metrics that were used for the purpose of the model validation are the correlation coefficient (r), the bias (BIAS), the mean absolute error (MAE), the root mean square error (RMSE), the scatter index (SI) and the symmetrical mean absolute percentage error (SMAPE). The definition of all metrics can be found in [20] apart from SMAPE, which is defined as follows:

$$\text{SMAPE} = \frac{200}{N} \sum_{i=1}^{N} \left| \frac{u_{\text{B},i} - u_{\text{M},i}}{u_{\text{B},i} + u_{\text{M},i}} \right| \tag{1}$$

where $u_{\text{B},i}$ is the $i-$th measured wind speed from the buoy, $u_{\text{M},i}$ is the $i-$th wind speed from the

model and N is the sample size. The corresponding values are bounded to $[0, 200]$. The model performance is characterized as good if all the values of the adopted metrics are as close to zero as possible except for correlation coefficient that should be close to unity. The statistical metrics that were used for the purpose of the model validation as regards wind direction are the circular-circular correlation coefficient (r_c), the bias (BIAS), the mean circular absolute error (MCAE), and the root mean error (RME); for the definition of these measures see [28]. If the values of BIAS, MCAE and RME are close to zero and r_c is close to unity, then the model performance is good. See Section 3, where the wind data from both data sources are presented in detail along with the results of the validation procedure.

After the model validation, the spatial distribution of offshore wind speed and direction and wind power density evaluated at 80 m height asl is provided in the annual and the seasonal time scale.

For the presentation of the main wind climate and wind power potential features for the Aegean and Ionian Seas various statistical quantities are assessed. As regards wind speed, apart from the standard statistical quantities, i.e. mean value and standard deviation, the mean annual variability (MAV) and the inter-annual variability (IAV) are estimated and presented herewith. These measures describe the variability of the variable of interest within a year and from one year to another, respectively, and constitute essential design parameters in wind energy applications (e.g. OWF planning). Their mathematical expressions are provided in [20]. Regarding wind direction, the description of the main statistical quantities (i.e. mean value, circular variance, circular standard deviation), which are less known, can be found in Section II of Soukissian et al. [20] and standard reference books regarding circular statistics, such as [29] and [30]; see also [21], where wind (and wave) directions are modelled and examined in detail. In this work the circular variance instead of the circular standard deviation, is preferred since the former is bounded in $[0,1]$, in contrast to the latter, which ranges within $[0,\infty)$.

The original time series of wind speed and direction that were extracted from the NWP model and analyzed are referred to 10 m height above sea level (asl). For the wind power estimation, the wind speed extrapolation at height of, say, h m asl (i.e. to a typical turbine hub height) is based on the log-law:

$$u_h = u_{10} \frac{\ln(h/z_0)}{\ln(10/z_0)} \tag{2}$$

where u_h is the wind speed at the examined height h asl, u_{10} is the wind speed at 10 m height asl (obtained from the NWP model results) and z_0 is the roughness length that equals to 0.001 m for open sea areas [31]. The above equation has been compared with wind fields from LiDAR measurements, revealing its good performance [32].

Furthermore, the mean wind power density $\overline{P}$ (W/m^2) is estimated directly by utilizing the long time series of wind speed at the model grid points. For a particular grid point, $\overline{P}$ is estimated by the following equation:

$$\overline{P} = \frac{1}{2N} \sum_{i=1}^{N} \rho u_i^3 \tag{3}$$

where N is the total sample size of the time series, ρ is the air density, considered to be constant and equal to 1.2258 kg/m^3, and u_i is the value of wind speed obtained from the corresponding time series. Without doubt, the accurate description of the wind climate in an area is a rather delicate issue, since rather small variations of u (from the actual value of u, if wind speed measurements were available) lead to very large variations in $\overline{P}$ due to the third power. The main variability characteristics of the above-mentioned quantities are also provided.

An additional factor that should be taken into account for the development of OWFs is the percentage of time that the wind speed is within the operational wind speed limits of any industrial offshore wind turbine. The majority of wind turbines are characterized by typical cut-in speed ~ 4 m/s and cut-out speed ~ 25 m/s. The examined event is denoted by $A = [4 \leq U_{80} \leq 25]$, where U_{80} denotes the wind speed at 80 m height asl, representing a typical turbine hub height. The probability of occurrence of A can be estimated from the following relation:

$$\Pr[A] \approx \frac{\sum_{i=1}^{N} \mathbf{1}_A \left(4 \leq u_{80,i} \leq 25 \right)}{N} \tag{4}$$

where $\mathbf{1}_A(\cdot)$ is the indicator function, $u_{80,i}$ is the particular $i-$th wind speed value (at 80 m height asl) and N is the considered sample size.

Finally, for some particular locations of the Aegean and Ionian Seas, the main wind climate and wind power density characteristics are also evaluated.

3.   Wind Data Description and Validation

3.1. General

The hindcast wind data spanning 15-year period (1995–2009) was produced by downscaling European Centre for Medium-Range Forecasts (ECMWF) global reanalyses with the aid of the Eta-based NWP model of the POSEIDON system [33]. In Table 1, the main features of the model are presented. Details about the dynamical downscaling procedure, the relevant products and the data evaluation can be found in the work of Papadopoulos et al. [34].

Table 1. Main features of the ETA model configuration.

Main features of the ETA model configuration	
Horizontal grid increment	0.10×0.10
Vertical levels	38
Time steps (s)	30
Grid	Arakawa E grid
Model output frequency (h)	3
Cloud microphysics scheme	Ferrier [35]
Cumulus scheme	Betts-Miller-Janjic [36]
Surface layer scheme	Monin-Obukhov-Janjic scheme [37]
Land-surface model	4-layer NOAH [38]
Planetary boundary layer scheme	Mellor-Yamada-Janjic Level 2.5 [37]
Radiation scheme (shortwave; longwave)	GFDL [39,40]

In order to evaluate the model results, measured wind speed and wind direction data obtained from four offshore oceanographic buoys of the POSEIDON system, see [41], were used; see Table 2, where the geographical coordinates of each buoy are depicted. In the next subsection, the statistical analysis of the collocated wind data sets from the NWP model and the buoy measurements is presented, and then, the validation of the model results is performed.

Table 2. Wind speed statistics for the buoy measurements and model data in the Aegean and Ionian Seas.

Location name	Latitude (deg)	Longitude (deg)	Data source	N	m (m/s)	s (m/s)	min (m/s)	max (m/s)	CV (%)
Athos	39.96	24.72	Buoy	19971	5.185	3.585	0.000	23.098	69.130
			Model		4.731	2.880	0.046	16.578	60.874
Lesvos	39.15	25.81	Buoy	20436	6.811	3.711	0.000	24.998	54.490
			Model		5.482	2.800	0.123	16.156	51.071
Mykonos	37.51	25.46	Buoy	20824	7.623	4.014	0.001	21.306	52.656
			Model		6.372	3.110	0.161	16.168	48.811
Pylos	36.83	21.61	Buoy	5072	5.280	2.941	0.270	18.065	55.710
			Model		4.622	2.658	0.094	14.769	57.510

3.2. Analysis and validation of collocated measured and model wind data

A typical procedure for a rational comparative assessment of buoy measurements and model results is to collocate wind speed data from both data sources in space and time. The spatial collocation is achieved by interpolating the four nearest model grid points to the exact buoy location via the implementation of the inverse squared distance weighting interpolation function to the simulated wind speed time series, while the temporal collocation is based on the simultaneous available wind speed data and refers to the common time steps 00, 03, ..., 21 UTC. Furthermore, outlying observations were detected using Cook's distance and discarded from the collocated samples after thorough examination.

In Table 2, the main statistics (i.e. mean value m, standard deviation s, minimum and maximum values: min, max, and coefficient of variation CV) of the collocated wind speed data are presented. The sample sizes N correspond to a consecutive wind time series of two to seven year-long. The values of m, s, max and CV are consistently higher for the buoy measurements (except for CV of Pylos), compared to the model.

In Table 3, the main directional statistics parameters (i.e. mean value $\overline{\theta}_W$ and circular variance V_θ) of the collocated wind direction data are presented. Mean wind direction for all selected locations is confined to the northern sector (roughly from NW to NE for both data sources). The smallest value of circular variance is exhibited for Mykonos, while the variance values for the model results are consistently higher, except for Pylos.

Table 3. Wind direction statistics for the buoy measurements and model data in the Aegean and Ionian Seas.

Location name	Data source	N	$\overline{\theta}_W$ (deg)	V_θ (-)
Athos	Buoy	19971	36.268	0.692
	Model		14.160	0.699
Lesvos	Buoy	20436	23.232	0.596
	Model		7.000	0.625
Mykonos	Buoy	20824	323.013	0.567
	Model		324.169	0.597
Pylos	Buoy	5072	334.119	0.722
	Model		306.537	0.690

In Table 4, the obtained results for each statistical metric are presented for the four selected buoy locations. From the values of r, it is evident that the linear relationship between buoy and model wind speed data is strong with the highest value observed for Mykonos (0.874) and the lowest one for Pylos (0.796). The absolute values of BIAS range from 0.455 m/s (corresponding to Athos) to 1.329 m/s (Lesvos) while the lowest (1.922 m/s) and the highest (2.563 m/s) values of RMSE are provided by Pylos and Lesvos. SMAPE, which is more sensitive when the model underestimates wind speed with respect to buoy measurements, takes the minimum value (35.264%) for Mykonos; the same holds true for SI (0.308).

Table 4. Statistical metrics applied for the validation of wind speed data from the model with reference to buoy measurements in the Aegean and Ionian Seas. Boldface numbers denote the best value for each metric.

Location name	r (-)	BIAS (m/s)	RMSE (m/s)	MAE (m/s)	SI (-)	SMAPE (%)
Athos	0.832	**0.455**	2.041	1.614	0.394	45.505
Lesvos	0.809	1.329	2.563	2.081	0.376	42.565
Mykonos	**0.874**	1.251	2.351	1.909	0.308	**35.264**
Pylos	0.796	0.658	**1.922**	**1.552**	**0.364**	40.082

Table 5. Statistical metrics applied for the validation of wind direction data from the model with reference to buoy measurements in the Aegean and Ionian Seas. Boldface numbers denote the best value for each metric.

Location name	r_c (-)	BIAS (deg)	MCAE (deg)	RME (-)
Athos	0.539	−22.108	38.247	0.472
Lesvos	0.639	−16.232	29.315	0.381
Mykonos	**0.708**	**1.156**	**22.706**	**0.330**
Pylos	0.447	−27.583	33.340	0.428

In Table 5 the obtained results for each statistical metric are presented for the four selected buoy locations. From the values of r_c, it is evident that the (circular) association between buoy and model wind direction data is fair with the highest value observed for Mykonos (0.708) and the lowest one for Pylos (0.447). The values of the other adopted metrics are consistently lower also for Mykonos.

In the work of Kotroni et al. [18], the performance of another high-resolution model was evaluated with reference to six onshore observing stations based on four evaluation statistics and the results were presented both for the complete and operational datasets, referring though only to wind speed (see Table 2 of [18]). Comparing the values of the common adopted metrics, i.e. r, RMSE and MAE, for the complete datasets, it seems that the performance of the present model is better since the lower values of RMSE and MAE, and the highest one of r suggest more accurate estimates of the wind field. Let us note though, that such comparisons should be made with caution since they refer to different reference measurements and different lengths of collocated wind data.

4.   Numerical Results

In this section, the obtained results by applying the above methodology are presented. The color gradation for the following figures, used for representing the spatial distribution of offshore wind speed and wind power density, at the different time scales was kept the same for facilitating comparison purposes while the location of the main names that are mentioned throughout the analysis are shown in Figure 8.

4.1. Wind speed and direction

The mean annual offshore wind speed and direction (at 10 m height asl) are presented in Figure 2a. It is noticed that the winds blow in the mean from the north-eastern directions for the N. Aegean Sea, while the north-western directions are the most characteristic over the Ionian Sea. The maximum value of the mean annual wind speed (at 10 m height asl) is observed in the central Aegean Sea (37.5°N, 25.6°E) reaching the value of 7.57 m/s with associated mean wind direction 327°. A gross comparison of the Aegean and Ionian Seas reveals that the Aegean is clearly windier. In Figure 2b, the standard deviation of the mean annual wind speed is depicted. The maximum standard deviation (0.38 m/s) is observed again at the central Aegean Sea (37.8°N, 25.3°E).

The rest panels of Figure 2 exhibit the seasonal spatial distribution of wind speed and direction. Summer wind speeds are characterized by the highest values with maximum value around 9 m/s over the Karpathian Sea (35.5°N, 27°E) and mean wind direction 304°, then winter follows with maximum value up to approximately 8 m/s over the N. Aegean Sea (39.3°N, 25.4°E) with mean wind direction 50°, then autumn with maximum value up to 7.35 m/s over the central part of the Aegean Sea (37.5°N, 25.6°E) and mean wind direction 329°, and finally, spring with maximum value up to 7.03 m/s over the E. Aegean Sea (37.7°N, 26.5°E) and mean wind direction 303°. Let us note though, that regarding the entire area of interest (i.e. the Aegean and Ionian Seas) winter wind speeds are, in the mean, clearly higher than summer wind speeds.

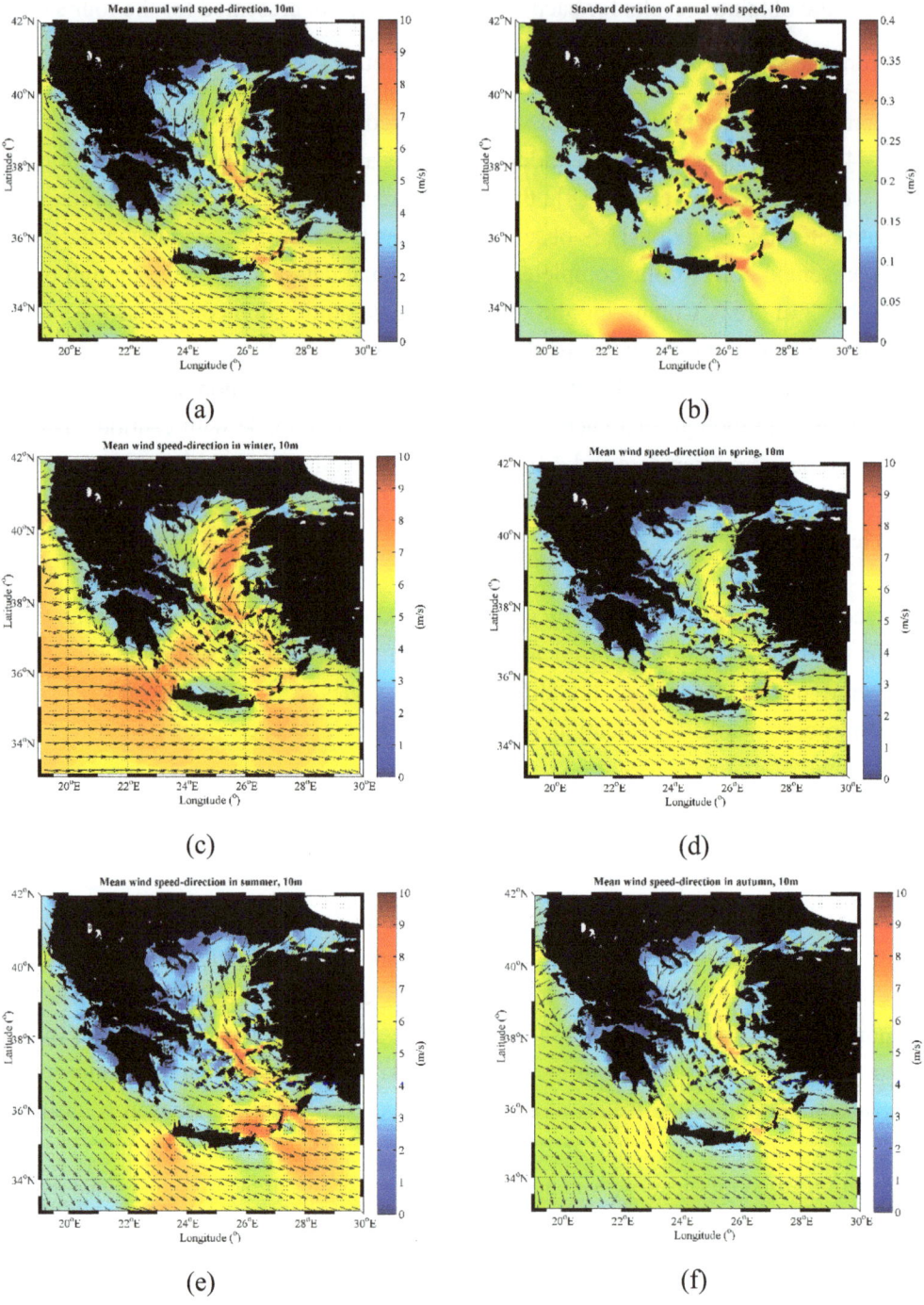

Figure 2. Spatial distribution of (a) mean annual offshore wind speed and wind direction, (b) standard deviation of the mean annual wind speed, (c) mean offshore wind speed and direction in winter, (d) spring, (e) summer and (f) autumn, at 10 m height asl in the Aegean and Ionian Seas for the period 1995–2009.

From the monthly analysis (not presented here) the overall windiest month for the Aegean Sea is February and then, December and January follow. However, the largest monthly values of wind speed are observed during July and August. Specifically, the largest mean monthly value (9.80 m/s) is observed during July over the Karpathian Sea (35.5°N, 27°E), and the second largest (9.0 m/s) during August at the same area. This distinct seasonality feature is clearly revealed when compared with the typical wind seasonality pattern observed at another location in the Ionian Sea. See, for example, Figure 3, where the seasonality curves of both mean monthly wind speed and direction for two selected locations in the Karpathian and Ionian Seas (37.5°N, 21°E) are depicted.

From Figure 3 it is evident that a remarkable stability of wind direction is present as regards the location of the Karpathian Sea. The mean monthly wind directions are totally restricted in the sector [270°, 305°], suggesting a rather unidirectional wind pattern. The most intense months as regards wind speed in this area are July, August and June. On the other hand, for the location of the Ionian Sea, the monthly wind speed curve follows the classical U shape; however, the wind direction exhibits a large fluctuation between 200° and 360°, except for December where the mean wind direction is around 60°.

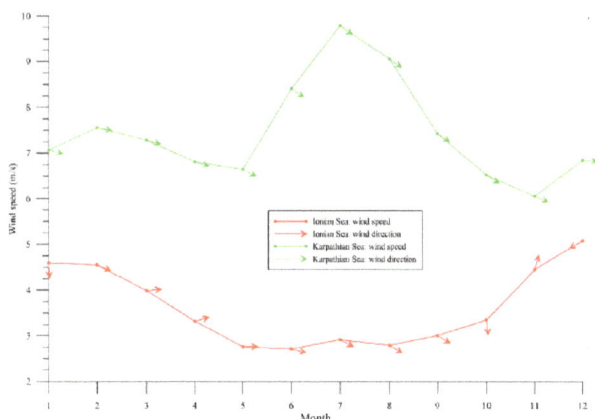

Figure 3. Seasonality of wind speed and wind direction at two characteristic areas of the Aegean and Ionian Seas.

From the above-mentioned results, the determinant role of the etesian winds, which blow over the Aegean Sea during June, July, August and September, is verified as regards the formation of the offshore wind climate of the area. The etesian winds shape to a large extent the corresponding wind climate and contribute significantly to the rich wind potential of the area.

The mean annual variability (MAV) and the inter-annual variability (IAV) of wind speed are depicted in Figure 4. The highest values of MAV and IAV are observed over the southern coasts of Peloponnesus (36.8°N, 22.2°E) with values ~ 88% and ~ 8%, respectively. The corresponding lowest values are observed over the south-western part (34.9°N, 23.1°E) and the northern part (36.8°N, 22.2°E) of Crete Isl., with values ~ 41% and ~ 2%, respectively. Clearly, areas with values of MAV and IAV as low as possible, combined with high mean values of wind speed, are favorable for the development of OWFs. Finally, in Figure 4c, the spatial distribution of the circular variance of the annual wind direction is shown. For the offshore areas, the corresponding values are very low (close to zero), while for some particular coastal areas, especially in the N. Ionian Sea, the circular variance may reach values up to 0.4.

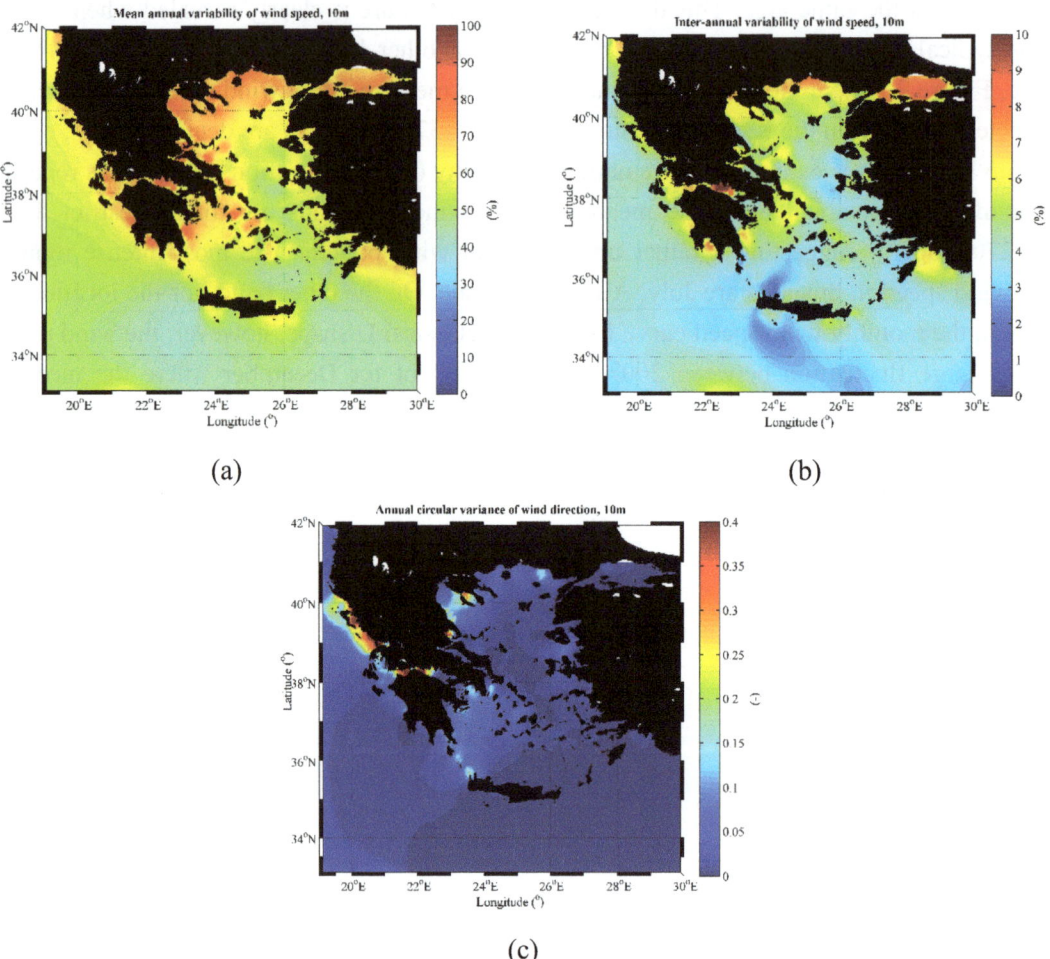

Figure 4. Spatial distribution of (a) mean annual variability, (b) inter-annual variability of offshore wind speed at 10 m height asl and (c) circular variance of annual wind direction in the Aegean and Ionian Seas for the period 1995–2009.

4.2. Wind power density

In Figure 5, the offshore wind power density is presented at 80 m height asl for the annual and seasonal scale. The overall highest value of the mean annual wind power density (~ 885 W/m^2) is depicted in the central Aegean Sea (37.7°N, 26.5°E). The behavior of wind power density at the seasonal scale can be described as follows: the overall highest value is observed during summer reaching peak values around 1172 W/m^2 over the south-eastern Aegean Sea (35.5°N, 27°E); winter follows with highest value ~ 1090 W/m^2 over the N. Aegean Sea (39.3°N, 25.4°E), then autumn with peak value ~ 806 W/m^2 over the central part of the Aegean Sea, and finally, spring with peak value ~ 773 W/m^2 over the E. Aegean Sea (37.7°N, 26.5°E).

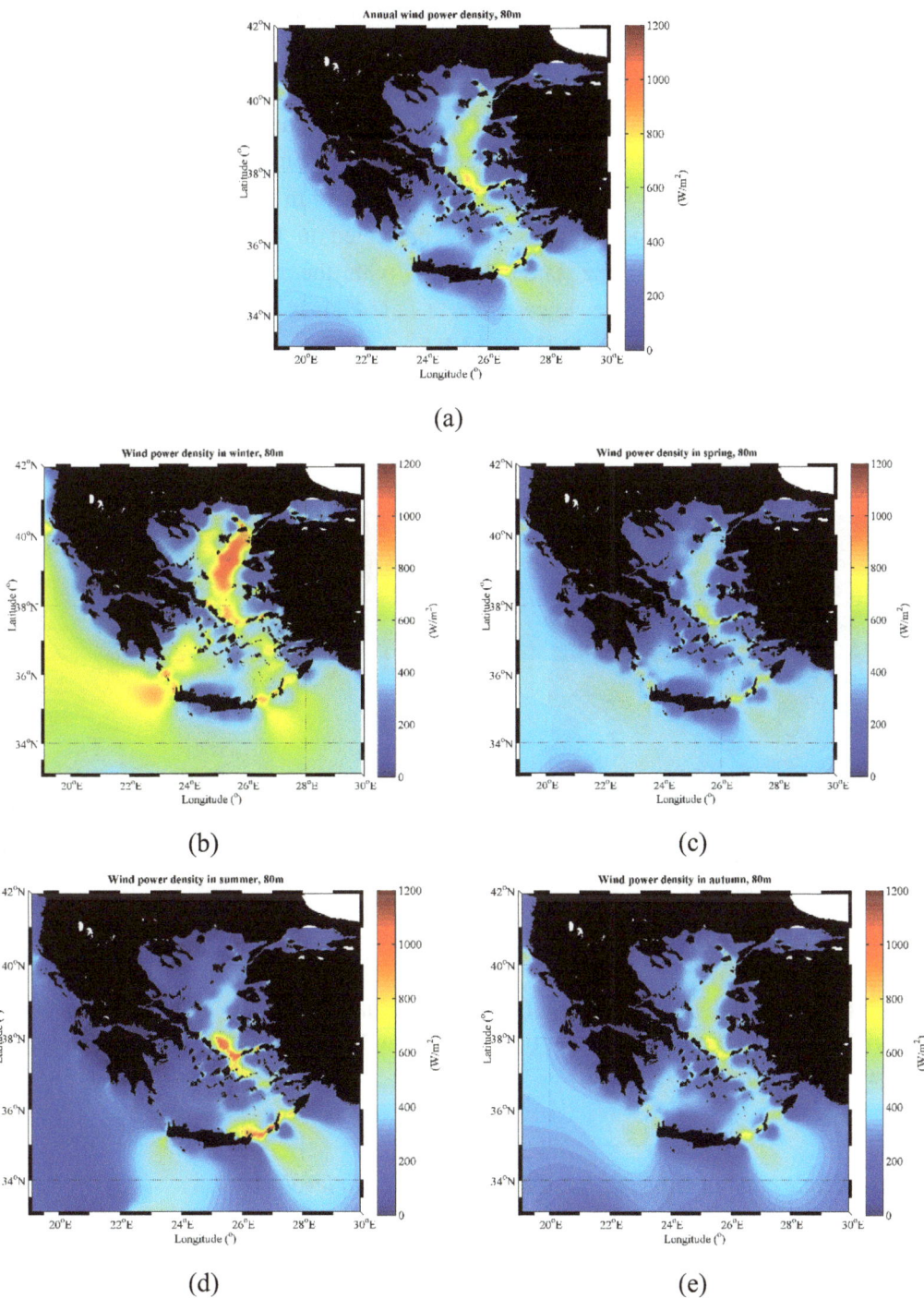

(a)

(b)

(c)

(d)

(e)

Figure 5. Spatial distribution of (a) mean annual offshore wind power density, (b) mean offshore wind power density in winter, (c) spring, (d) summer and (e) autumn, at 80 m height asl in the Aegean and Ionian Seas for the period 1995–2009.

In the Ionian Sea, the wind power density reaches values up to ~ 584 W/m^2 over the northern part (40.2°N, 19.3°E) and ~ 490 W/m^2 over the southern part (35.5°N, 22°E). During winter, the corresponding maximum wind power density values are ~ 790 W/m^2 and ~ 780 W/m^2. Overall, the lowest values of wind power density in the Ionian Sea are observed during summer.

The spatial distribution of offshore wind power potential resembles qualitatively the results of Kotroni et al. [18] (see Figure 3 of [18]), stressing that the Aegean Sea is clearly more energetic than the Ionian Sea with values higher than 600 W/m^2 at 50 m height asl. The same holds true for the results of a higher resolution atmospheric model presented in Emmanouil et al. [19] (see Figure 5 of [19]), where offshore wind power potential is presented at 10 m height asl.

MAV and IAV of wind power density are presented in Figure 6. For MAV, the highest value (~ 375%) is depicted at the same location with the corresponding measure for wind speed (36.8°N, 22.2°E) while the lowest one (~ 114%) is very close (spatially) with the corresponding location for wind speed (33°N, 23.1°E). The highest value of IAV (~ 30%) is located over the W. Aegean Sea (38.8°N, 23.8°E) and the lowest one (~ 6%) over the northern part of Crete Isl. (34.8°N, 23.7°E).

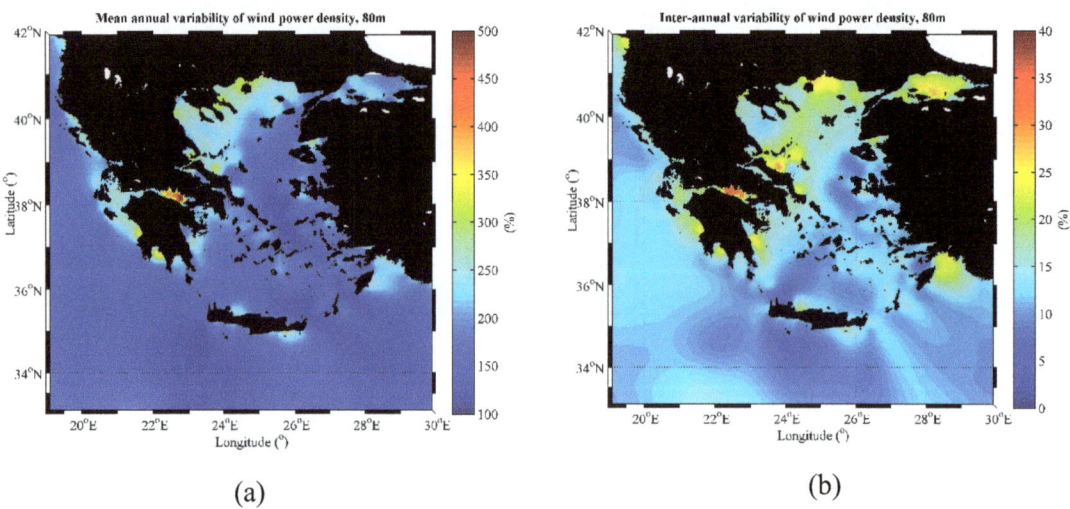

Figure 6. Spatial distribution of (a) mean annual variability and (b) inter-annual variability of offshore wind power density at 80 m height asl in the Aegean and Ionian Seas for the period 1995–2009.

4.3. Equiprobable contours of wind speed at 80 m above sea level

In Figure 7, the spatial distribution of the frequency of the event $\left[4\,\text{m/s} \leq U_{80} \leq 25\,\text{m/s}\right]$ is depicted. As can be seen in this figure extended areas of the central Aegean Sea, the S. Karpathian Sea and the area at the southwestern part of Crete Isl., are characterized by frequencies of occurrences close to 0.9. The corresponding seasonal features (not presented here) suggest that the persistence of wind speeds within the wind turbine operational limits in the southern part of the Aegean Sea takes its maximum values during summer. On the other hand, autumn seems to be the period of the year with the smallest persistence of the examined wind speeds. For the Ionian Sea, the highest persistence is observed at the northern and offshore southern part.

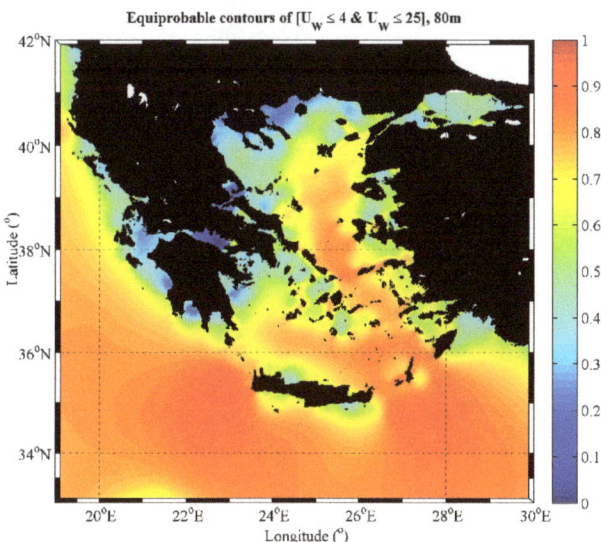

Figure 7. Equiprobable contours of the event [4 m/s $\leq U_W \leq$ 25 m/s] at 80 m height asl in the Aegean and Ionian Seas for the period 1995–2009.

4.4. Wind climate and wind power density characteristics for specific locations in the Aegean and Ionian Seas

In this section, the main statistical characteristics of wind climate and wind power density are provided for seven locations where OWF development is at the concept/planning phase. Six of these locations cover the entire Aegean Sea and one is located in the N. Ionian Sea; see also Figure 8. In Table 6, the mean annual wind speed ($\bar{U}_{80}$), the corresponding standard deviation ($s_{U_{80}}$) along with MAV and IAV, and the annual mean wind power density ($\bar{P}_{80}$) at 80 m height asl are provided. As is seen in this table, the location in Kasos (Karpathian Sea) is the most favorable as regards wind resource availability; the mean annual wind speed is 8.03 m/s, the values of MAV (49.83%) and IAV (4.09%) are the lowest amongst the examined locations and the available wind potential is 570.9 W/m². The second best location is Steno Kafirea with an annual wind speed of 7.51 m/s, a rather high value of IAV (5.51%) while the available wind potential is 545.6 W/m². The lowest wind availability is exhibited at the location in the Thrace Sea (with mean annual wind speed at 3.33 m/s); the same location exhibits also the highest value of MAV (72.55%) and IAV (6.09%). Overall, it seems that the available wind resource characteristics at Kasos location are superior rendering the location favorable for a future potential OWF development.

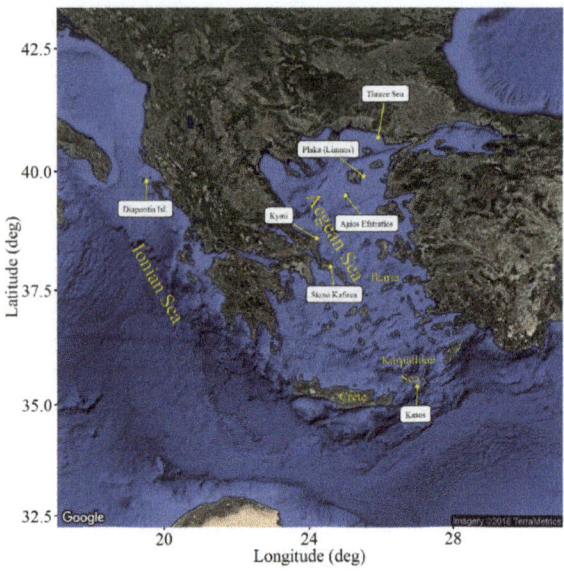

Figure 8. Locations of the main sea areas and islands mentioned in the text. The location names inside a box denote the seven locations that were selected for the detailed analysis of wind climate and wind power density characteristics (the background of the map has been derived from Google Earth).

Table 6. Wind climate and wind power density characteristics of selected locations in the Aegean and Ionian Seas.

Location name	Latitude (deg)	Longitude (deg)	$\overline{U}_{W,80} - s_{W,80}$ (m/s)	MAV (%)	IAV (%)	$\overline{P}_{80}$ (W/m²)
Agios Efstratios	39.5	25.0	6.941 – 0.336	59.75	4.84	468.2
Diapontia Isl.	39.8	19.5	5.089 – 0.222	62.09	4.36	191.4
Kymi	38.6	24.2	6.620 – 0.319	58.77	4.81	383.4
Kasos	35.4	27.0	8.029 – 0.328	49.83	4.09	570.9
Thrace Sea	40.7	25.9	3.339 – 0.203	72.55	6.09	71.7
Plaka (Limnos)	39.9	25.5	6.692 – 0.319	61.38	4.77	430.3
Steno Kafirea	38.0	24.6	7.513 – 0.414	57.34	5.51	545.6

It is important to remind that except for wind resource availability other important criteria should be also considered such as bottom depth, distance to shore and to the power grid, type of sediments, distance to ports, etc. On the other hand, socio-economic and environmental considerations may alter significantly the final decisions as regards the feasibility of OWFs in any location. Finally, since the results presented in this work are based on the analysis of hindcast data, they are always subject to numerical model uncertainties as is already mentioned in Section 2.

5.   Conclusions

In this study, the offshore wind power potential of the Aegean and Ionian Seas is analytically examined and presented in order to rationalize and facilitate the preliminary identification of promising candidate sites for future offshore wind farm development. Since offshore *in situ* measurements are scarce and cannot cover the spatial scale of the examined sea areas, the analysis performed was based on a 15-year hindcast of wind data spanning from 1995 to 2009. The hindcast data was produced by a high-resolution atmospheric downscaling of the ECWMF reanalysis, exhibiting fine scale features in the coastal zone and allowing the analysis even close to the land/sea boundary. An analytic validation of the model results with respect to offshore *in situ* measurements obtained from four oceanographic buoys was implemented, confirming its suitability for the purpose of this study. Regarding the Greek Seas, validation of wind direction, apart from wind speed, is presented here for the first time as far as the authors are aware of. The validation results suggest that the model performance is overall good: as regards wind speed, the correlation coefficient ranges between 0.874 and 0.796, the absolute BIAS between 0.455 m/s and 1.329 m/s and RMSE between 1.922 m/s and 2.563 m/s; as regards wind direction, the circular-circular correlation coefficient ranges between 0.708 and 0.447, the absolute BIAS between 1.156 deg and 27.583 deg and the circular absolute error between 22.706 deg and 38.247 deg.

From the statistical analysis of offshore wind speed and direction (at 10 m height asl) it can be highlighted that the etesian winds are the most distinct wind feature for the Aegean Sea, blowing mainly during summer, when the wind conditions of the rest Mediterranean Sea are considerably calmer while, simultaneously, energy demand tends to be higher due to tourism and recreation activities. Moreover, variability measures referring to wind speed, wind direction and wind power potential as well as persistence statistics for wind speed that are important for the offshore wind energy community are provided. Overall, it seems that the wind potential (at 80 m height asl) mainly in the Aegean Sea and, secondarily, in the Ionian Sea, is adequately exploitable at specific locations; offshore deep water locations, especially in the Aegean Sea, are characterized by high values of offshore wind resource (mean annual wind power density up to 885 W/m^2 in the central Aegean). The overall highest value is observed during summer reaching peak values around 1172 W/m^2 over the south-eastern Aegean Sea. In addition, the persistence of wind speeds within the wind turbine operational limits in the Aegean Sea is very suitable for offshore wind energy projects since extended areas of the Aegean Sea are characterized by frequencies of occurrences close to 0.9.

However, some of the examined shallow-water locations (see Section 4.4) do not seem to meet the required standards for the efficient exploitation of offshore wind resource. On the other hand, expansion of the offshore wind energy industry in deep water areas requires a significant progress towards commercialization of floating wind turbines and reduction of costs.

The presented results are expected to give insight to the offshore wind energy sector as regards wind energy potential estimates and variability measures in the Aegean and Ionian Seas while such analysis can form the basis of a preliminary assessment study for the potential exploitation of offshore wind in the future. Furthermore, wave energy assessment studies and meteorological applications can be benefited from relevant analyses. Let us though point out that the local in-depth assessment of any candidate area is the only robust means to provide final conclusions as regards the suitability of the area from all considered frameworks (i.e. technical/engineering, socio-economic and environmental aspects). For example, a detailed wind resource assessment, requires at least 2-year *in situ* wind measurements for the most accurate estimation of the wind power potential in a candidate area for OWF development.

At the end of the day, it remains to be seen whether the new law 4414/2016 for RES will provide the impetus needed for the development of the offshore wind energy industry in Greece.

Acknowledgments

This research has been funded from the Greek General Secretariat for Research and Technology and the European Regional Development Fund under Grant Agreement no. 09SYN-32-598 for the project "National programme for the utilization of offshore wind potential in the Aegean Sea: preparatory actions" (AVRA).

Conflict of Interest

All authors declare no conflicts of interest in this paper.

References

1. European Wind Energy Association, The European offshore wind industry-key trends and statistics 2015. European Wind Energy Association, 2016. Available from: https://www.ewea.org/fileadmin/files/library/publications/statistics/EWEA-European-Offshore-Statistics- 2015.pdf.

2. Bilgili M, Yasar A, Simsek E (2011) Offshore wind power development in Europe and its comparison with onshore counterpart. *Renew Sust Energ Rev* 15: 905–915.

3. Perveen R, Kishor N, Mohanty SR (2014) Off-shore wind farm development: present status and challenges. *Renew Sust Energ Rev* 29: 780–792.

4. Soukissian TH, Papadopoulos A (2015) Effects of different wind data sources in offshore wind power assessment. *Renew Energ* 77: 101–114.

5. Colmenar SA, Perera PJ, Borge DD, et al. (2016) Offshore wind energy: a review of the current status, challenges and future development in Spain. *Renew Sust Energ Rev* 64: 1–18.

6. European Wind Energy Association, Wind in power: 2015 European statistics. European Wind Energy Association, 2016. Available from: http://www.ewea.org/fileadmin/files/library/ publications/ statistics/ EWEA-Annual-Statistics-2015.pdf.

7. Soukissian T, Reizopoulou S, Drakopoulou P, et al. (2016) Greening offshore wind with the smart wind chart evaluation tool. *Web Ecol* 16: 73–80.

8. Kaldellis JK, Apostolou D, Kapsali M, et al. (2016) Environmental and social footprint of offshore wind energy. Comparison with onshore counterpart. *Renew Energ* 92: 543–556.

9. Brownlee MTJ, Hallo JC, Jodice LW, et al. (2015) Attitudes toward offshore wind energy development. *J Leisure Res* 47: 263–284.

10. Westerberg V, Jacobsen JB, Lifran R (2013) The case for offshore wind farms, artificial reefs and sustainable tourism in the French Mediterranean. *Tourism Manag* 34: 172–183.

11. DIRM Méditerranée, Document de planification: Le développement de l'éolien en mer Méditerranée. France: Préfecture maritime de la Méditerranée, Préfecture de région Provence Alpes Côte d'Azur, 2015. Available from: http://www.dirm.mediterranee.developpement-durable.gouv.fr/ IMG/pdf/ Document_de_planification_pour_transmission.pdf.

12. 4C Offshore, Two more French Floaters get approved! 4C Offshore, 2016. Available from: http://www.4coffshore.com/windfarms/two-more-french-floaters-get-approved!-nid4813.html.

13. Rodrigues S, Restrepo C, Kontos E, et al. (2015) Trends of offshore wind projects. *Renew Sust Energ Rev* 49: 1114–1135.

14. Westerberg V, Jacobsen JB, Lifran R (2015) Offshore wind farms in Southern Europe-determining tourist preference and social acceptance. *Energ Res Soc Sci* 10: 165–179.

15. Zountouridou EI, Kiokes GC, Chakalis S, et al. (2015) Offshore floating wind parks in the deep waters of Mediterranean Sea. *Renew Sust Energ Rev* 51: 433–448.

16. Onea F, Deleanu L, Rusu L, et al. (2016) Evaluation of the wind energy potential along the Mediterranean Sea coasts. *Energ Explor Exploit* 34: 766–792.

17. Balog I, Ruti PM, Tobin I, et al. (2016) A numerical approach for planning offshore wind farms from regional to local scales over the Mediterranean. *Renew Energ* 85: 395–405.

18. Kotroni V, Lagouvardos K, Lykoudis S (2014) High-resolution model-based wind atlas for Greece. *Renew Sust Energ Rev* 30: 479–489.

19. Emmanouil G, Galanis G, Kalogeri C, et al. (2016) 10-year high resolution study of wind, sea waves and wave energy assessment in the Greek offshore areas. *Renew Energ* 90: 399–419.

20. Soukissian T, Karathanasi F, Axaopoulos P (2017) Satellite-based offshore wind resource assessment in the Mediterranean Sea. *IEEE J Oceanic Eng* 42: 73–86.

21. Soukissian TH (2014) Probabilistic modeling of directional and linear characteristics of wind and sea states. *Ocean Eng* 91: 91–110.

22. Song M, Chen K, Zhang X, et al. (2016) Optimization of wind turbine micro-siting for reducing the sensitivity of power generation to wind direction. *Renew Energ* 85: 57–65.

23. Watson SJ (2014) Quantifying the variability of wind energy. *Wires Energ Environ* 3: 330–342.

24. EMODnet, EMODnet Bathymetry portal. EMODnet, 2016. Available from: http://www.emodnet-hydrography.eu/.

25. Caralis G, Chaviaropoulos P, Ruiz Albacete V, et al. (2016) Lessons learnt from the evaluation of the feed-in tariff scheme for offshore wind farms in Greece using a Monte Carlo approach. *J Wind Eng Ind Aerod* 157: 63–75.

26. Greek Parliament, Governmental Gazette, A' No. 149/9-8-2016, L. 4414/2016. Official Government Gazette of the Hellenic Republic.

27. European Commission, Official Journal of the European Union, Guidelines on State aid for environmental protection and energy 2014-2020 (2014/C 200/01). European Commission, 2014. Available from: http://eur-lex.europa.eu/legal-content/EN/TXT/PDF/?uri=CELEX:52014XC 0628(01)&from=EN.

28. Karathanasi FE, Soukissian TH, Axaopoulos PG (2016) Calibration of wind directions in the Mediterranean Sea. In: Proceedings of the 26th International Ocean and Polar Engineering Conference; 2016; Rhodes, Greece, 491–497.

29. Fisher N (1995) Statistical analysis of circular data. 1st ed. Cambridge: Cambridge University Press, 294.

30. Jammalamadaka R, SenGupta A (2001) Topics in circular statistics. Singapore: World Scientific Publishing Co. Pte. Ltd., 334.

31. Hansen FV (1993) Surface roughness lengths. White Sands Missile Range, New Mexico: U.S. Army Research Laboratory, 1–40.

32. Shu ZR, Li QS, He YC, et al. (2016) Observations of offshore wind characteristics by Doppler-LiDAR for wind energy applications. *Appl Energ* 169: 150–163.

33. Papadopoulos A, Katsafados P (2009) Verification of operational weather forecasts from the POSEIDON system across the Eastern Mediterranean. *Nat Hazards Earth Syst Sci* 9: 1299–1306.

34. Papadopoulos A, Korres G, Katsafados P, et al. (2011) Dynamic downscaling of the ERA-40 data using a mesoscale meteorological model. *Mediterranean Mar Sci* 12: 183–198.

35. Ferrier BS, Jin Y, Lin Y, et al. (2002) Implementation of a new grid-scale cloud and precipitation scheme in the NCEP Eta Model. 19th Conference on weather analysis and forecasting/15th Conference on numerical weather prediction. San Antonio: Am Meteorol Soc, 280–283.

36. Janjic ZI, Gerrity JP, Nickovic S (2001) An alternative approach to nonhydrostatic modeling. *Mon Weather Rev* 129: 1164–1178.

37. Janjić ZI (1994) The step-mountain Eta coordinate model: further developments of the convection, viscous sublayer, and turbulence closure schemes. *Mon Weather Rev* 122.

38. Chen F, Janjić Z, Mitchell K (1997) Impact of atmospheric surface-layer parameterizations in the new land-surface scheme of the NCEP mesoscale eta model. *Bound Lay Meteorol* 85: 391–421.

39. Lacis AA, Hansen J (1974) A parameterization for the absorption of solar radiation in the earth's atmosphere. *J Atmos Sci* 31: 118–133.

40. Schwarzkopf MD, Fels SB (1991) The simplified exchange method revisited: an accurate, rapid method for computation of infrared cooling rates and fluxes. *J Geophys Res* 96: 9075–9096.

41. Soukissian T, Chronis G (2000) Poseidon: a marine environmental monitoring, forecasting and information system for the Greek Seas. *Mediterranean Mar Sci* 1: 71–78.

8

Influence of design parameters on the structural and fatigue behaviors of a floating point wave energy converter

Pedro J. B. F. N. Beirão [1,2,]* **Cândida M. S. P. Malça** [1], and **Raimundo P. Felismina** [1]

[1] Instituto Politécnico de Coimbra, ISEC, DEM, Coimbra, Portugal
[2] LAETA, IDMEC, Instituto Superior Técnico, Universidade de Lisboa, Lisboa, Portugal

* **Correspondence:** Email: pbeirao@isec.pt

Abstract: The demand for electricity production has been consistently raising since the last century. In the future, the tendency is to grow even further. Concerning this fact, renewable energy and specifically, wave energy should be considered as an alternative for energy production. However, devices suitable to harness this renewable energy source and convert it into electricity are not yet commercially competitive. This paper is focused on the structural analysis of a wave energy converter (WEC) through the numerical study of several design parameters. Tridimensional computer aided design (3D CAD) numerical models were built and several Finite Element Analyses (FEA) were performed using a commercial finite element code. The main components of the WEC were simulated assuming different materials. The Von Mises stress gradients and displacement fields determined by FEA demonstrated that, regardless of the WEC component, materials with low Young's modulus seems to be unsuitable for this application. The same is valid for the material yield strength since materials with higher yield strength lead to a better structural behavior of the WEC components. The developed 3D CAD numerical model showed to be suitable to analyze different combinations of structural conditions.

Keywords: wave energy converter; structural analysis; finite element analysis; failure

1. Introduction

The demand for energy production, specifically for electricity, has been consistently raising since the last century. More recently, in the last decades the world energy consumption increased considerably and in the future the tendency is to grow even further. Renewable energy sources will soon play an important role. Among them, ocean wave energy is one of the most promising alternatives to fossil fuels regarding the production of electricity [1]. The energy that can be extracted from sea waves offers a high-power density when compared with more developed technologies like solar and wind energies. Although suffering from strong inter-annual and inter-seasonal variabilities [2,3], wave energy supplies a more stable source of renewable energy, when compared with those renewable energies. This allows the utilization of wave energy converters (WEC) over the year. Moreover, their customization through engineering solutions will permit to match those devices to different ocean climates [4].

Although at an early stage of development, wave energy must be considered as a promising source of energy supply. However, suitable devices to deal with this kind of energy source and convert it into electricity are not yet technically feasible and commercially available. It is expected that they will be in the medium and long term, as they are currently under development. Some of the several WEC concepts being tested have already reached full scale [5]. However, they are not yet competitive and require a great deal of investigation in several areas, namely from structural, dynamics and aerodynamics standpoints. Finite element analysis (FEA) has been extensively used in several engineering domains. In the renewable energy field, the structural and fatigue analysis is mainly dedicated to wind systems [6,7,8]. Still, its application to the wave energy field is not so much exploited, being primarily devoted to mooring systems [9] and inspired by the knowledge gathered from offshore floating structures, like oil platforms and shipbuilding [10,11,12].

This numerical study is focused on structural and fatigue analysis of a floating point WEC through numerical results obtained by FEA. The objective is to avoid plastic deformation and short fatigue life. It is divided in several parts, described as follows: a) The first purpose comprises the analysis of several WEC design parameters, such as the type of materials and dimensions employed in the WEC components. It will be shown if there is any relation between different materials and dimensions, and stress concentrations and displacement fields, to avoid plastic deformation. b) The second goal includes the analysis of the influence of several buoy characteristics, such as dimensions, geometries and submerged conditions. It will be shown if there is any relation between buoy dimensions or components' dimensions or the selection of different materials and stress concentrations and displacement fields, to avoid plastic deformation. In what concerns the buoy geometries, it will be determined which is the one leading to the best WEC structural behavior. Concerning the buoy submerged conditions, it will be determined the buoy critical position. c) The final aim comprises the analysis between fatigue and materials and dimensions selected for the WEC components.

It is expected that numerical results, obtained by means of simulations, could give some guidance to improve the structural components, to enhance their mechanical performance on the structural and fatigue behaviors of the entire WEC. Specifically, it is intended to show if there is any relation between applied forces (simulating several sea conditions) and the selection of dimensions and materials of the WEC components. WEC components can, thus, be optimized to reach the best structural performance, according with several specific characteristics [13–16].

Finally, this paper does not aim to show any kind of correlation between structural problems and energy production. Moreover, this paper is not focused on any structural problems observed in a specific WEC.

2.   Wave Energy Converter

2.1. Characterization

Based on a well proven conventional design of a near shore floating point absorber WEC defined by a very small characteristic dimension when compared with the typical wave length [17] this study analyzes, from a structural standpoint, several design parameters. The main components of the WEC used in this study include a floating buoy (assumed to have different characteristics) connected by supporting cables to a double effect hydraulic cylinder. An electric linear generator, connected to the piston rod of the double effect hydraulic cylinder acts as power take-off (PTO) responsible by the conversion of the wave energy into electrical energy. A cardan joint connects the double effect hydraulic cylinder to the mooring system. The cardan joint allows six modes of motion, however and due to simplicity reasons the buoy is assumed to oscillate only in heave mode. Figure 1 schematizes the WEC used in this study.

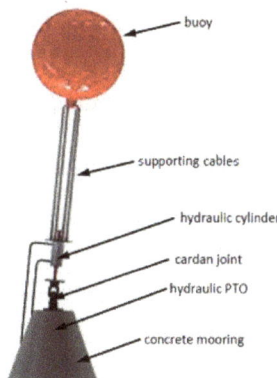

Figure 1. 3D CAD model of the WEC.

Its working principle is quite simple: when submitted to the sea waves the buoy heaves upwards under the influence of a wave crest and heaves downwards under the effect of a wave trough. Therefore, the piston rod of the hydraulic cylinder can travel between a fully advanced position and a fully retracted position, respectively.

2.2. Numerical model

The objective of this study is to analyze the influence of several design parameters, such as buoy characteristics–dimensions, geometries and submerged conditions–on the WEC structural behavior, when a resultant force is applied to the external surface of the buoy. Therefore, three different geometries were considered for the buoy–spherical, cylindrical and tulip (tulip geometry results from a combination between a cone and a cylinder). Additionally, for the spherical buoy geometry, three different radiuses were assumed. Three different submerged conditions were also considered for the buoy–totally submerged, partially submerged and buoy at the surface.

To perform the study, a tridimensional computer aided design (3D CAD) numerical model was initially built and several FEA were performed through the Simulation tool of SolidWorks®

software [18]. This commercial finite element code needs, among other inputs, the pressure value related with the resultant force that acts upon the buoy. That value was obtained, by means of simulations, by a time domain simulator obtained from the dynamic model of the WEC [19]. Figure 2 shows the boundary and load conditions applied to the developed 3D CAD models [20].

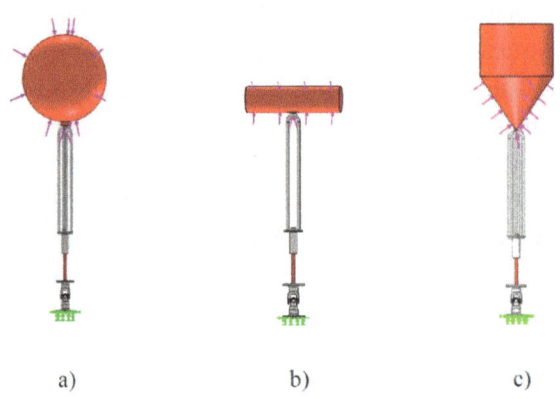

a)   b)   c)

Figure 2. 3D CAD models with boundary and load conditions.

Concerning the boundary conditions, it can be observed that the inferior half of the cardan joint is rigidly fixed. This means that constraints of no displacements and rotations are applied to simulate the WEC mooring system at the sea bottom [20].

The finite element model for the three buoy geometries is shown in Figure 3. For a better accuracy, a relatively fine mesh of triangular elements was applied. The selected solid mesh resulted from the meshing sensitivity study previously performed [20].

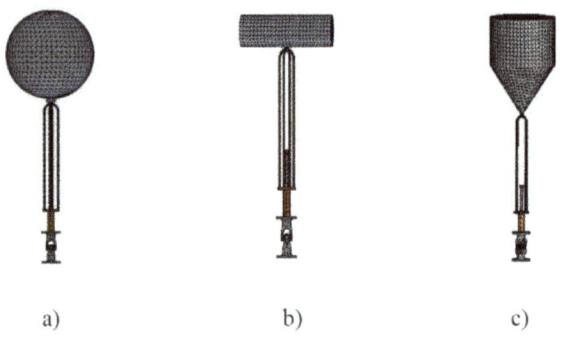

a)   b)   c)

Figure 3. Mesh geometry for the three buoy geometries.

Regarding the elastic material properties of the buoy, most of the buoys commercially available employ different materials for the core and shell. However, due to simplicity reasons, it was assumed that both the buoy core and shell are made of polyethylene. Additionally, it is assumed that all remaining components of the WEC are made of stainless steel AISI 316 (SS) and/or high strength steel (HSS). Table 1 summarizes the relevant elastic material properties considered for both the buoy and the remaining components of the WEC [20].

Table 1. Selected elastic material properties for the WEC main components.

Material	Young's modulus [MPa]	Poisson coefficient	Yield strength [MPa]	Density [kgm^{-3}]
Polyethylene	1860×10^{-6}	0.39	30	940
AISI 316 SS	193000	0.27	172	8000
HSS	21000	0.28	620	7700

3. Dynamic Model and Simulator

The resultant external force assumed to act upon the buoy was computed by means of simulations obtained from a dynamic simulator based on the dynamic model of the WEC. Due to the dynamic model nonlinearities, the simulator was built in the time domain using Matlab/Simulink software.

The dynamic model of the WEC assumes that the floating buoy is excited by sea waves. Linear wave theory is used since it is also assumed that wave amplitudes and oscillations are sufficiently small when compared with the wavelength [5,17]. The external force acting on the buoy results from the influence of two forces: i) hydrostatic force resulting from the combination of PTO force, buoyancy force exerted on the buoy due to the instantaneous buoy position with respect to the seawater free surface and drag force; ii) hydrodynamic force decomposed into two components acting upon the wetted buoy surface, known as heave excitation force due to the incident waves acting upon the assumed stationary buoy and radiation force due to the energy transfer from the heaving buoy to the waves that are being radiated away from the buoy [17]. The derived equations were grouped into a dynamic model block, under individual subsystems. Another block simulates the sea waves. Several inputs such as buoy and wave data, among others are needed to run the simulation. A more detailed description about the dynamic model and simulator can be found in [19].

Pressure values, needed as inputs in the commercial finite element code, were obtained from the relation between the resultant external force acting upon the buoy [it was assumed a maximum resultant hydrodynamic force of 50×10^3 N (peak to peak amplitude)] and the projected area of the buoy. Those values can change due to variations, for instance, in the buoy geometry or radius and/or in wave parameters. If the wave input parameters are modified different results will be obtained.

4. Results and Discussion

To analyze the influence of several buoy characteristics–dimensions, geometries and submerged conditions–on the structural behavior of the WEC, several simulations were done using the Simulation tool of SolidWorks® software [18]. FEA provides an insight into the magnitude of the displacement (deformation) and stress concentrations, as well as their location. In the following figures areas where maximum values occur are displayed in red. Notice that when considering the wave crest, the hydraulic cylinder piston rod is assumed to be at fully advanced position. This position of the hydraulic cylinder piston rod leads to a higher magnitude of stresses in the piston rod than those achieved for the retracted position of the hydraulic cylinder piston rod, as shown in [20]. Therefore, results presented hereafter correspond to the full advanced hydraulic cylinder piston rod.

4.1. Influence of buoy submerged conditions

The influence of the three different submerged buoy conditions–buoy at the surface, partially and totally submerged buoy–on the structural behavior of the WEC is shown in Figures 4 and 5, respectively. Stress and displacement fields are displayed for the spherical buoy geometry.

For the buoy at the surface, a considerable decrease on both maximum stresses and displacements was obtained. This is remarkable for the totally submerged buoy, where stresses and displacements reached values around six and eight times, respectively, lower than those verified for the partially submerged condition of the buoy. Among the three positions considered, the partially submerged buoy position corresponds to the critical buoy position. Moreover, this is the most expected position of the buoy when it heaves due to the action of the sea waves. Therefore, only the submerged condition corresponding to the buoy partially submerged was considered in the following analysis.

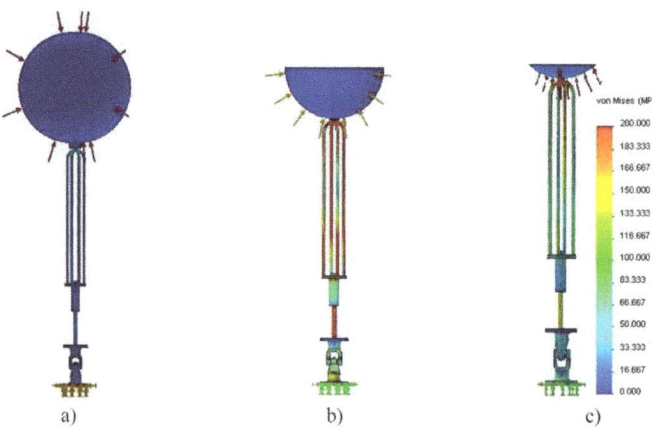

Figure 4. Von Mises stress field for 200 mm radius spherical buoy for different buoy submerged conditions: a) totally submerged, b) partially submerged and c) at the surface.

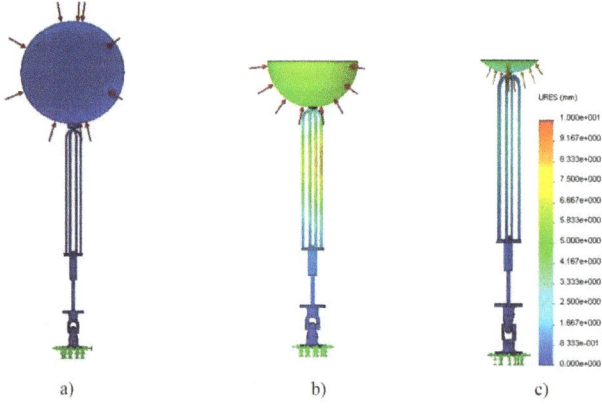

Figure 5. Displacement field for 200 mm radius spherical buoy for different buoy submerged conditions: a) totally submerged, b) partially submerged and c) at the surface.

4.2. Influence of buoy dimensions

Figure 6 shows the Von Mises stress gradient determined by FEM for partially submerged polyethylene spherical buoys with radius of 100 mm, 200 mm and 400 mm, respectively. Corresponding displacement fields are shown in Figure 7.

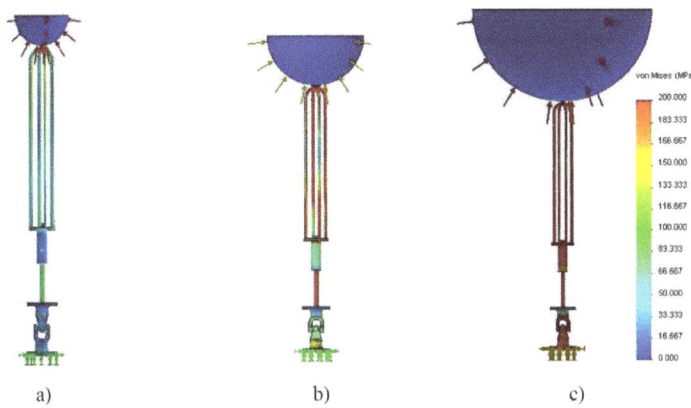

a)   b)   c)

Figure 6. Von Mises stress field for a) 100 mm, b) 200 mm and c) 400 mm radius spherical.

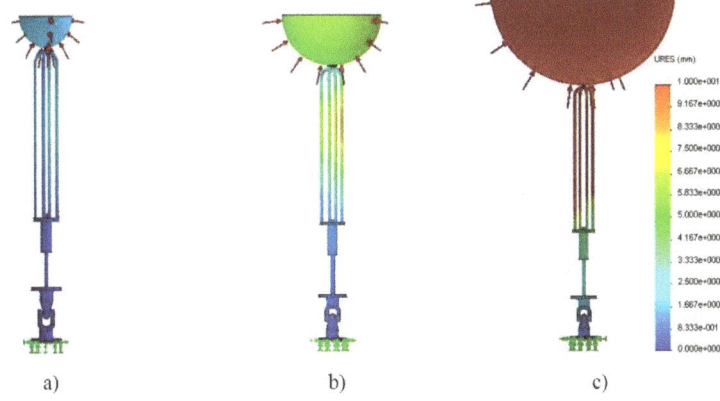

a)   b)   c)

Figure 7. Displacement field for a) 100 mm, b) 200 mm and c) 400 mm radius spherical.

It can be seen from Figures 6 and 7 that the spherical buoy with radius of 100 mm leads to the lowest values of stresses and displacements and, consequently, to the best mechanical structural behavior because the yield strength of the materials is never attained and plastic deformation is not reached.

For the radius of 200 mm and 400 mm the scenario is substantially different. Plastic deformation is reached for most of the components (excluding the buoy) since maximum stress values largely exceed the yield strength of AISI 316 SS. The collapse of the structure is, therefore, predictable. This conclusion is supported by the displacement field values shown in Figure 7. The increase of the buoy dimensions lead to a significantly increase of the displacements, reaching unacceptable values for the two highest radii simulated. As expected, maximum displacements take place at the buoy and at the supporting cables, whereas areas of highest stress concentration occur in the piston rod and in the supporting cables.

4.3. Influence of buoy geometries

The influence of the buoy geometry on the structural behavior was also studied. Figures 8 and 9 illustrate the resulting Von Mises stress and displacement fields for the three different buoy geometries under study.

Results reveal that for maximum stress values, no significant differences were observed between geometries. However, for the cylindrical buoy geometry, a greater number of WEC components are submitted to maximum stress values. Additionally, since maximum stresses are greater than the yield stress of the AISI 316 SS used for the supporting cables, piston rod and cardan joint, plastic deformation is reached. The greatest magnitude of stresses and displacements founded in the cylindrical geometry is justified by its higher projected area (where the corresponding pressure is applied).

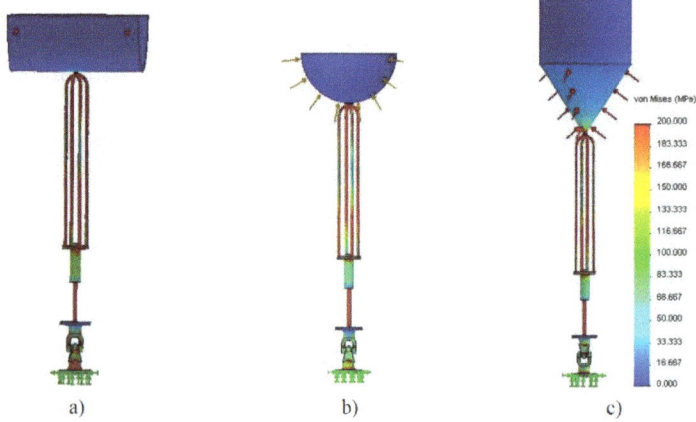

Figure 8. Von Mises stress field for a) cylindrical, b) 200 mm radius spherical and c) tulip.

If the comparison between buoy geometries is made in terms of the displacement field values, the superior mechanical behavior achieved by the spherical geometry is unquestionable, which is confirmed by the lower displacements obtained, as shown in Figure 9. For cylindrical and tulip buoy geometries and for the design parameters considered–applied load, dimensions of the components and materials–the displacements have the same magnitude of the component dimensions, which means that greater deformations take place and that the collapse of the structure is reached.

This situation can be avoided by adjusting (function of the magnitude of the applied resultant force) the dimensions and/or through a suitable selection of the elastic material properties of the WEC components, as shown in the following section.

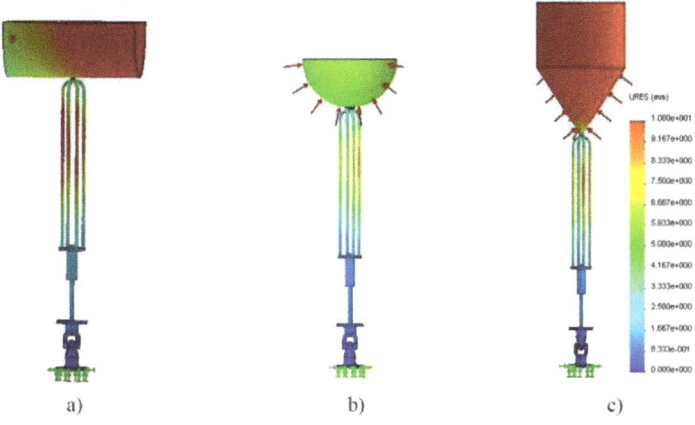

Figure 9. Displacement field for a) cylindrical, b) 200 mm radius spherical and c) tulip.

4.4. Influence of the material and dimensions selected for the WEC components

The WEC numerical model that was developed allows the single simulation of each one of the following design parameters–material and dimensions selected for the WEC components or their combination.

As an example, Figure 10 shows, for the 200 mm radius spherical buoy, the influence on maximum stress and displacement values if HSS with yield strength of 620 MPa is selected for the components, instead of the original SS with yield strength of 172 MPa (values from Table 1). Although a slight increase on the buoy deformation, maximum stresses decrease substantially.

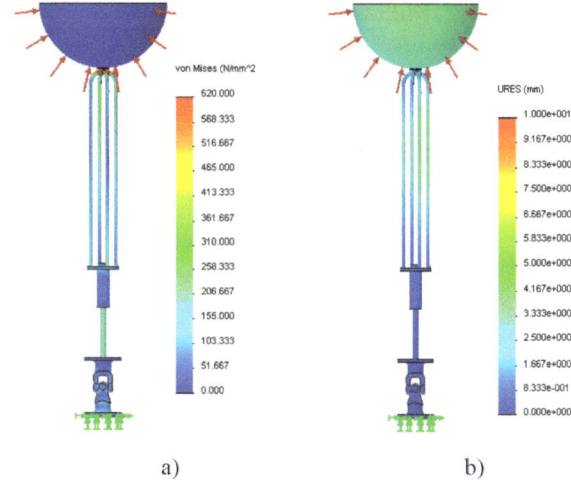

Figure 10. a) Von Mises stress and b) displacement fields for 200 mm radius spherical buoy, HSS components, 15 mm supporting cables and 20 mm piston rod diameters.

Concerning the resizing of the WEC, an increase of 5 mm on both the supporting cables and piston rod diameters was considered. Additionally, a different selection of materials was also assumed for the components, which are now assumed to be made of HSS instead of SS.

Figure 11 shows the results obtained in terms of maximum stress and displacements. A significant reduction of maximum stress and displacement values was obtained.

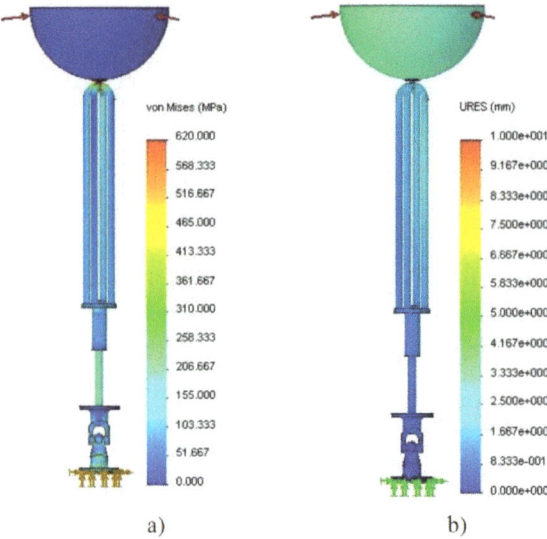

a)   b)

Figure 11. a) Von Mises stress and b) displacement fields for 200 mm radius spherical buoy, HSS components, 20 mm supporting cables and 25 mm piston rod diameters.

Table 2 summarizes all cases and combinations previously studied.

Table 2. Studied cases and combinations.

	Von Mises stress field [MPa]	Displacement field [mm]
Cylindrical	Figure 8	Figure 9
Spherical	Figures 4, 6, 8	Figures 5, 7, 9
Tulip	Figure 8	Figure 9
Buoy totally submerged	Figure 4	Figure 5
Buoy partially submerged	Figure 4	Figure 5
Buoy at the surface	Figure 4	Figure 5
100 mm	Figure 6	Figure 7
200 mm	Figures 4, 6, 8, 10, 11	Figures 5, 7, 9, 10, 11
400 mm	Figure 6	Figure 7

Table 3 compiles the maximum Von Mises stress field and displacement field values, numerically obtained by means simulations.

From the obtained FEA results, simulated load magnitude, dimensions of the components and selected material, it is shown that: i) increasing the buoy dimensions leads to a significant increase of both maximum values of stresses and displacements, which means that an increase of plastic deformation can be expected for the highest radius tested; ii) spherical buoy geometry leads to the lowest values of stresses and displacements, when compared with the tulip, but mainly with cylindrical buoy geometries and iii) partially submerged condition corresponds to the critical buoy position since highest maximum stresses and displacements are reached.

Table 3. Maximum values.

Figure	Von Mises stress field [MPa]	Figure	Displacement field [mm]
Figure 4a	33	Figure 5a	0.9
Figure 4b	194	Figure 5b	7.6
Figure 4c	176	Figure 5c	3.4
Figure 6a	137	Figure 7a	2.7
Figure 6b	194	Figure 7b	7.6
Figure 6c	198	Figure 7c	9.7
Figure 8a	198	Figure 9a	9.5
Figure 8b	194	Figure 9b	7.6
Figure 8c	197	Figure 9c	9.7
Figure 10a	370	Figure 10b	5.1
Figure 11a	230	Figure 11b	3.4

4.5. WEC fatigue lifetime prediction

The influence of the materials selected for the components and corresponding dimensions, as well as the increase of stress concentration on the WEC fatigue lifetime estimation is briefly presented in this section.

The fatigue lifetime of the WEC components is predicted by the number of cycles to failure N. The Soderberg failure criterion was chosen to perform FEA [21]. A load cycle with a stress ratio R of −1 was chosen to simulate the external resultant force acting upon the buoy due to the sea waves.

The simultaneous action of fatigue and corrosion causes pitting at the surface of the components. Pitting corrosion acts as stress concentrations points, where fatigue cracking will start and grow rapidly. Since steels do not exhibit a definite endurance limit when tested in a corrosive environment, the surface finish factor, K_A, which represents the surface condition modification factor, could be taken as a fatigue strength reduction factor ranging from 0.1 to 1 [22,23]. This factor was used to simulate numerically the action of the corrosion since it is equivalent to the reduction of the number of cycles that cause failure at a certain alternating stress [24]. Values of K_A equal to 0.1, which simulates an accelerated corrosion process and 0.5, corresponding to an acceptable surface finish value for a considerable number of cycles in the long-term fatigue, were chosen to simulate the fatigue phenomenon with and without corrosion, respectively [23].

The influence of two different fatigue strength reduction factor K_A on the fatigue life prediction is shown in Figure 12.

An effective reduction in fatigue life occurs with the decreasing of K_A due to the increase of the induced stress concentration, as shown in Figure 12a. In fact, as better surface finish is, lower is the stress surface concentration. This means that the component fatigue strength is greater because the probability of fatigue nucleation cracks is reduced.

Nevertheless, for a number of cycles greater than 10^6, surface parameters such as surface finish, type of coating, surface material metallurgical state, surface residual stresses, among others, must be improved.

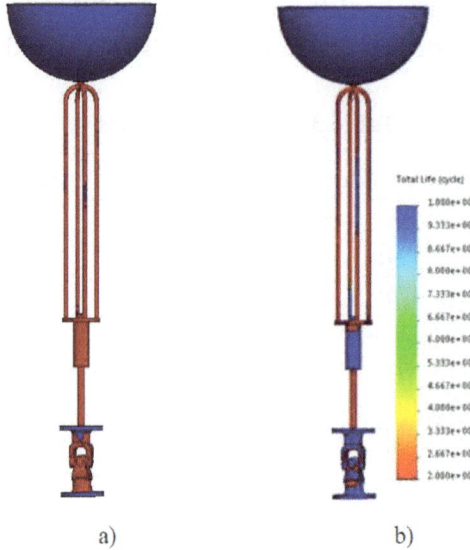

Figure 12. Number of cycles to failure for 200 mm radius spherical buoy, 15 mm supporting cables, 20 mm piston rod diameter, HSS components: a) $K_A = 0.1$ and b) $K_A = 0.5$.

Alternatively, other kind of materials or components dimensions can be tested, as shown through Figures 13 and 14.

The influence of two different materials on the fatigue life prediction is shown in Figure 13. As expected, SS presents a better fatigue strength when compared with HSS. However, both materials lead to a reduced number of cycles before failure namely at the supporting cables and piston rod.

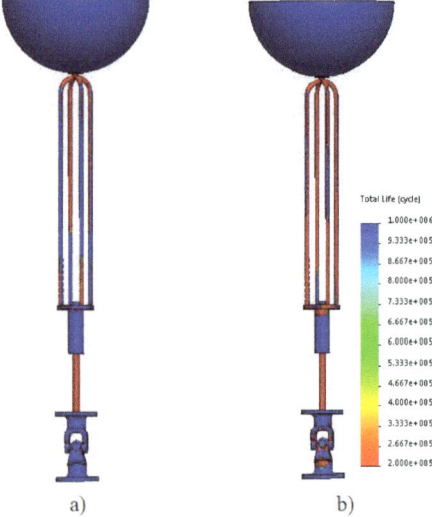

Figure 13. Number of cycles to failure for 200 mm radius spherical buoy, 15 mm supporting cables, 20 mm piston rod diameter, $K_A = 0.5$ for a) SS and b) HSS components.

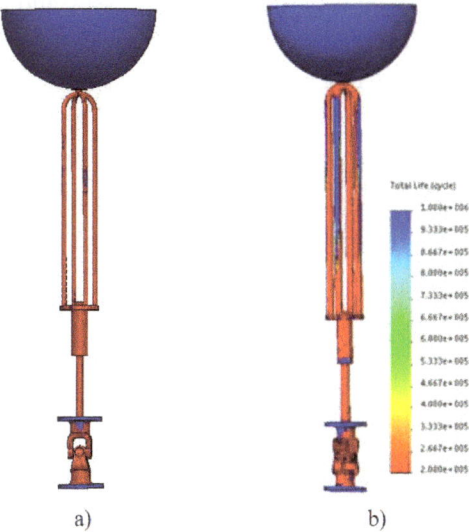

a)   b)

Figure 14. Number of cycles to failure for 200 mm radius spherical buoy, HSS components, $K_A = 0.1$ and for a) 15 mm supporting cables and 20 mm piston rod diameter, and b) 20 mm supporting cables and 25 mm piston rod diameter.

Concerning the influence of the dimension of the components on the fatigue strength, as expected and shown in Figure 14, no effective reduction in fatigue life is observed. Figure 14b shows a very slight increase of fatigue strength mainly for the supporting cables. When corrosion and fatigue happen simultaneously it can be expected that the increase on the dimensions of the components affect positively the fatigue strength. In fact, in fatigue with corrosion the crack initiation phase is very small, becoming the duration of the propagation phase dominant in the component life. Thus, it can be expected that greater component dimensions lead to greater fatigue strength. This is due to the longest path the crack must travel through until the occurrence of the final break of the component.

From these results, it can be concluded that the developed numerical model can also be used to predict and optimize the fatigue lifetime of the WEC components.

5. Conclusion

A brief characterization of the WEC used in this study is initially given. A numerical model of the WEC was developed and its performance evaluated using a finite element commercial code. The influence of several design parameters, such as buoy dimensions, geometries and submerged conditions on the structural behavior of the WEC, was analyzed and numerical results were obtained. Supporting cables were assumed as rigid bodies, thus neglecting their flexibility. For the magnitude of the simulated applied force, dimensions of the components and materials selected, FEA results, obtained by means of simulations, demonstrate that the increase in buoy dimensions leads to the increase on both maximum stress concentrations and displacement fields. The increase in buoy dimensions requires the resizing of the components or the selection of materials with greater yield strength, to avoid plastic deformation. It was also shown that the spherical buoy geometry leads to the best WEC structural behavior since it induces lower stresses and displacements. Furthermore, it was also demonstrated that the cylindrical buoy geometry leads to a chaotic scenario due to a high level of stresses and displacements when compared with spherical and tulip geometries. In what concerns the three buoys submerged conditions, it was verified that the partially submerged buoy corresponds to the critical buoy position, since highest maximum stresses and displacements are reached when compared with the other conditions. Results obtained from fatigue life analysis demonstrate the capacity of the developed WEC numerical model to predict the maximum number of load cycles before failure.

Conflict of Interest

All authors declare no conflicts of interest in this paper.

References

1. Valério D, Beirão P, Sá da Costa J (2007) Optimization of wave energy extraction with the archimedes wave swing. *Ocean Eng* 34: 2330–2344.
2. Guillou N, Chapalain G (2015) Numerical modelling of nearshore wave energy resource in the sea of Iroise. *Renew Energ* 83: 942–953.
3. Neill S, Hashemi M (2013) Wave power variability over the northwest European shelf seas. *Appl Energ* 106: 31–46.
4. Leijon M, Danielsson O, Eriksson M, et al. (2006) An electrical approach to wave energy conversion. *Renew Energ* 31: 1309–1319.
5. Falcão A (2010) Wave energy utilization: a review of the technologies. *Renew Sust Energ Rev* 14: 899–918.
6. Shoele K, Prowell I, Zhu Q, et al. (2011) Dynamic and structural modelling of a floating wind turbine. *Int J Offshore Polar* 21: 155–160.
7. Robertson A, Jonkman J (2011) Loads analysis of several offshore floating wind turbine concepts, Proceedings of the 21th International Offshore and Polar Engineering Conference (ISOPE), Maui, Hawaii, USA.
8. Aubault A, Cermelli C, Roddier D (2009) WindFloat: a floating foundation for offshore wind turbines–Part III: Structural analysis, Proceedings of the 28th International Conference on Ocean, Offshore and Arctic Engineering, Honolulu, Hawaii, USA.
9. Kim B, Sung H, Kim J, et al. (2013) Comparison of linear spring and nonlinear FEM methods in dynamic coupled analysis of floating structure and mooring system. *J Fluid Struct* 42: 205–227.
10. Fujikubo M (2005) Structural analysis for the design of VLFS. *Mar Struct* 18: 201–226.
11. Dong P (2005) A robust structural stress method for fatigue analysis of offshore/marine structures. *J Offshore Mech Arct* 127: 68–74.
12. Hansen V, Wang L, Sodahl N, et al. (2004) Guidelines on coupled analyses of deepwater floating systems, Offshore Technology Conference, Houston, USA.
13. Thorburn K, Karlsson KE, Wolfbrandt A, et al. (2006) Time stepping finite element analysis of a variable speed synchronous generator with rectifier. *Appl Energ* 83: 371–386.
14. Yueh CY, Chuang SH (2012) A boundary element model for a partially piston-type porous wave energy converter in gravity waves. *Eng Anal Bound Elem* 36: 658–664.

15. Nader JR, Zhu SP, Cooper P (2012) A finite-element study of the efficiency of arrays of oscillating water column wave energy converters. *Ocean Eng* 43: 72–81.
16. Wang S, Soares C (2014) Numerical study on the water impact of 3D bodies by an explicit finite element method. *Ocean Eng* 78: 73–88.
17. Falnes J (2004) Ocean waves and oscillating systems: linear interactions including wave-energy extraction, 1st Ed., Cambridge, Cambridge University Press.
18. Planchard D, Planchard M (2013) SolidWorks 2013 tutorial with video instruction, Schroff Development Corporation.
19. Beirão P, Valério D (2015) Numerical comparison between deep water and intermediate water depth expressions applied to a wave energy converter. *AIMS Energ* 3: 525–546.
20. Malça C, Beirão P, Felismina R (2014) Influence of material selection on the structural behavior of a wave energy converter. *AIMS Energ* 2: 359–372.
21. Shigley J, Mischke C (1988) Mechanical engineering design, 5th Ed., McGraw-Hill Inc. Available from https://www.amazon.com/Mechanical-Engineering-Design-McGraw-Hill/dp/0070568995.
22. Spotts M (1998) Design of machine elements, 7th Ed., Prentice Hall.
23. Campbell F (2008) Elements of metallurgy and engineering alloys, ASM International.
24. CATI Tech Notes: Fatigue Strength Reduction Factor, 2014. Available from: https://www.cati.com/blog/2014/11/fatigue-strength-reduction-factor/.

9

Global changes in total and wind electricity (1990–2014)

María del P. Pablo-Romero [1,]* **and Rafael Pozo-Barajas** [2]

[1] Department of Economic Analysis and Political Economy, Faculty of Economics and Business Sciences, University of Seville, Ramon y Cajal 1, 41018 Seville, Spain
[2] Department of Financial Economy and Operations Management, Faculty of Economics and Business Sciences, University of Seville, Ramon y Cajal 1, 41018 Seville, Spain

* **Correspondence:** Email: mpablorom@us.es

Abstract: Wind energy is one of the renewable energies which have less adverse environmental impact and is becoming economically affordable long before several other renewable energies. Over recent years, substantial additions have been noted in wind energy capacity, although many differences can be observed between countries. Using the latest available data, this paper provides a concise analysis of wind energy and electricity consumption trends for the period 1990–2014 in a dual perspective, by principal world regions and by per capita gross national income levels in 2014. Electricity consumption has been divided into three types of energy: non-renewable, renewable excluding wind and wind energy. Annual rates of change, energy intensity, energy in per capita terms and some ratios have been analyzed. Notable regional differences and trends are observed in the studied variables. The first 15 European Union countries, other developed countries (ODC) and East Asian (EAS) and South Asian countries (SAS) are the regions which currently have the highest wind capacity.

Keywords: climate and energy policy; wind electricity; worldwide regional analysis

JEL code: Q2, Q48, Q58, R1, R58

---

1. Introduction

The international political response to climate change began at the Rio Earth Summit in 1992. The adoption of the UN Framework on Climate Change (UNFCCC) was included in this convention, with the aim of stabilizing atmospheric concentrations of greenhouse gases (GHGs) to avoid dangerous interference with the climate. The UNFCCC, which entered into force in 1994, has been reviewed in the annual Conference of Parties (COP). Thus, the Kyoto Protocol was adopted in the COP3 and the Green Climate Fund was created in the COP17. In the COP21 (Paris Climate Conference), an action plan was set out in order to avoid climate change, by keeping the increase in global average temperature to well below 2 °C [1].

The COP21 [2] contains the steps all countries have the responsibility to make before 2020. Thus, each country shall prepare and communicate its nationally determined contributions (NDCs) every five years. Much work is to be done in this respect because, as stated in Climate Action Tracker [3], only five intended NDCs are already considered sufficient. Hence, as stated in Brauers and Richter [4], almost all countries have to scale-up their 2025–2030 contributions. Therefore, to achieve this target, energy and GHG emission must be reduced in all economic sectors.

In order to reduce these emissions, action on energy policy is essential, as emissions from the energy sector represent roughly two-thirds of all GHG emissions [5]. Therefore, these policies are essential for tackling the climate change problem. The main origin of emissions in the energy sector is the use of fossil fuels, as over 90% of energy-related emissions are CO_2 from fossil-fuel combustion. By sector, as stated in IEA [6], the largest emitting sector is power generation (40% in the world in 2014), mainly due to the use of fossil fuels. Therefore, reducing the use of fossil fuels is especially relevant within the power generation sector. In that regard, the use of renewable energies is one of the main measures to be applied and, over recent years, many countries have started to produce electricity and heat from renewable energies. Among renewable energies, wind energy is believed to have the least adverse environmental impact and is one of the renewable energies that has become economically affordable long before several others [7]. As a result, it is worth noting the substantial additions of wind capacity between 2000 and 2014, although many differences can be observed between countries [8].

Despite the importance of the increase in electricity consumption and wind capacity, there are few studies focusing on its heterogeneity across the world. Some studies refer to a specific part of the world. Thus, the study by Scarlat et al. [9] analyzes the bioenergy contribution in the European Union, the study by Dent [10] analyzes the wind energy development in East Asia and Europe, and the study by Mohammed et al. [11] presents the status of renewable energy consumption, including wind, in Sub-Saharan Africa. Likewise, some studies refer to specific countries, for example, the study by Chingulpitak and Wongwises [12] refers to wind energy in Taiwan, the study by Ho [13] refers to wind energy in Malaysia and the study by Kaplan [14] refers to Turkey. From a more general perspective, Leung and Yang [15] provide an overview of wind turbine development and development trends of offshore wind farms in some countries, such as China, the USA, Germany, Spain and India. However, very little consideration has been given to some regions such as Africa, Transition economies, Latin America and the Middle East. However, as stated in Mundaca et al. [16], the causes and impacts of climate change are mostly couched in regional terms.

In addition, although some studies, such as those by IEA [8,17], refer to the electricity trends, these studies do not provide a detailed regional vision. Although there are some previous studies referring to renewable energy trends [18,19], they also do not offer a detailed regional vision in terms of the main world regions or of income groups, nor do they specifically focus on wind energy. In addition, these studies do not analyze the electricity or wind electricity consumption related to economic and population growth. In that sense, few data have been previously provided about wind regional trends and its main economic indicators. Therefore, it is appropriate to develop much more research on these geographical differences [20,21].

Following these previous studies, the aim of this short paper is to provide a concise analysis of electricity and wind energy consumption trends for the period 1990–2014 in a dual perspective, by principal world regions, covering 11 regions, and by per capita gross national income (GNI) levels in 2014, covering 4 groups. Thus, the study focused on electricity and wind energy trends from a worldwide regional perspective in total consumption and per inhabitants and by GDP terms, for which, to our knowledge, there are no previous specific studies. Considering world regions by geographic and income groups allows an illustration as to what extent the electricity and wind energy consumption evolution is a geo-economic process. In that regard, the choice of the 11 regions and four groups of GNI levels are related to geographical and economic issues, in line with the main classifications provided by the World Bank [22] and the United Nations [23]. Alternative factors, such as atmospheric pressure and temperature, may also be adopted as criteria to categorize regions consuming electricity and wind energy, however this analysis goes beyond the objective of this work. The results of the paper may be taken as a starting point for additional and deeper analyses regarding regional economies.

2. Methodology

Wind electricity is analyzed by comparing its main indicators with respect to other general electricity sources. Therefore, electricity consumption (E) world data are analyzed and presented by world regions and by GNI levels. The E has been divided into three types of electricity sources: non-renewable electricity (NRE), renewable electricity excluding wind electricity (RE) and wind electricity (WE). These variables were analyzed by comparing annual changes, energy intensity (ratio of energy consumption to GDP in PPP) and energy in per capita terms. Additionally, the trend of the following ratios has been studied: total electricity consumption to total energy consumption, total renewable electricity to total electricity consumption and wind electricity to total renewable electricity consumption.

NRE is the electricity consumed which has been generated by using non-renewable energies. It is calculated as follows:

$$NRE = nre * Electricity\ consumption$$

$$nre = \frac{NRREE\ used\ in\ the\ Elec\ production}{Elec\ production + Elec\ imports}$$

where NRREE refers to non-renewables energies and Elec to electricity. NRREE is calculated as the sum of coal, oil, gas, nuclear and other sources used in the electricity generation.

RE is the electricity consumed which has been generated by using renewable energies, with the exception of wind. It is calculated as follows:

$$RE = re * Electricity\ consumption$$

$$re = \frac{RREE\ used\ in\ the\ Elec\ production}{Elec\ production + Elec\ imports}$$

where RREE refers to renewable energies and Elec to electricity. The RREE used in the electricity production is calculated as the sum of biofuels, waste, hydro, geothermal and solar PV, solar thermal and tide used in the electricity generation.

Finally, WE is the electricity consumed which has been generated by using wind energy. It is calculated as follows:

$$WE = w * Electricity\ consumption$$

$$w = \frac{wind\ energy\ used\ in\ the\ Elec\ production}{Elec\ production + Elec\ imports}$$

The world has been divided into the following regions: the first 15 members European Union (EU15), new European Union members (EU+), other developed countries (ODC), economies in transition (ET), Middle East and North African (MENA), East Asian (EAS), South Asian (SAS), Central African (CA), Southern African (SA), Latin American (LAC) and The Caribbean (CAR) countries. In addition, four groups have been considered by dividing the world according to GNI per capita in 2014: High income (HI), upper medium income (UMI), lower medium income (LMI), and, lower income (LI) countries (See Appendix).

The time period analyzed is 1990–2014, for which there are enough available data for all sample countries. Although the IEA [24] provide monthly electricity statistics, the last being for February 2017, this database only provides data on electricity production and not on its consumption. Also, it does not detail data on wind energy, but only on renewable energies as a whole. Therefore, data based on energy balances and heat and electricity production provided by IEA [25] have been used, the latest data for which refers to 2014.

3. Results

3.1. Total energy, electricity and wind electricity consumption trend in the period 1990–2014

Figure 1 shows world total energy and total and wind electricity consumption trends during the 1990–2014 period. It is worth noting the notable lower growth trend for electricity with respect to total energy consumption during the whole period. Likewise, it is noticed that total consumed electricity was mostly generated by non-renewable sources, although in recent years, the share of renewable sources has increased slightly. Thus, the non-renewable electricity consumption shows a similar growth trend with respect to that of electricity. It is also worth noting that wind electricity consumption shows growth from 2005, but it represents a small share of electricity consumption.

By regions, Table 1 shows that the electricity growth rate is nearly double the total energy growth rate, being 101% and 48%, respectively. Additionally, all regions have electricity growth rates higher than that of total energy, except for Southern Africa. The regions with highest electricity growth rates are East Asia (668%), South Asia (311%) and MENA (310%), while Central African, Latin American and The Caribbean regions show electricity growth rates higher than 100%. In absolute terms, East Asian and the other developed countries have the highest values. The East Asian region electricity consumption increase is 52% of the world electricity increase, while the other developed countries increase is 16%. The electricity consumption growth may be related to the economic growth in the world regions. Several studies have previously related electricity consumption to economic growth in a double sense [26]. Firstly, electricity consumption is considered necessary to generate growth, as are other energy sources such as fuels. Secondly, it is also seen as a consequence of this economic growth. In this latter sense, electricity consumption especially grows in the residential sector which may be linked to urbanization, lifestyle changes and the development of modern society. Thus, Ferguson et al. [27,28] show that while a well correlated relationship is observed between electricity use and wealth creation, there is no correlation between total energy use and wealth, indicating the essential role that electricity, rather than energy in general, plays in the development of modern society. Therefore, as regions grow, the electricity consumption may grow by a higher percentage than does total energy.

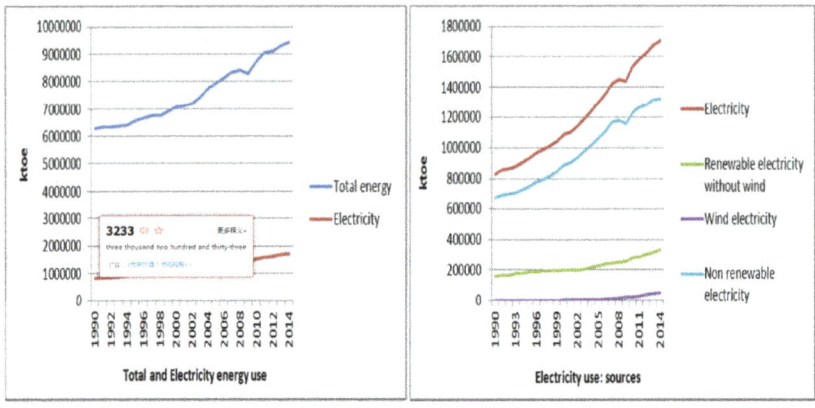

Figure 1. Total energy and total and wind electricity consumption.

Columns 5–7 in Table 1 show the participation in the total electricity consumption increase of renewable, wind and non-renewable electricity consumption. Globally, 18.65% of the electricity consumption increase was generated by using renewable electricity, while 5.2% was generated by using wind energy. It is worth noting that the highest renewable electricity consumption participations are observed in new European Union members, Latin American and EU15 countries, while the highest participation of wind electricity consumption is observed in the EU15 and new European Union members, with 29% and 24%, respectively. Likewise, the other developed countries region has a notable wind electricity participation in the electricity consumption increase, as 11% of this increase is generated by using wind energy. It can also be highlighted that the participation of wind in the electricity consumption increase in some regions has been very small, such as, for example, in Southern African, economies in transition, Central African and The Caribbean regions. In these regions, this participation is less than 1%. In line with these findings, the European Commission [29] already included the suggestion for a target share of renewable energy in 1997, with Denmark (1979), Portugal (1988), Germany (1989) and the United Kingdom (1989/90) being the first countries to introduce policy supports to promote renewable energy to produce electricity. Thus, prior to 2000, 15 countries in the EU had introduced policy measures to promote electricity from renewables and all of them have had policy support schemes since 2007 [30]. Regarding wind energy, it is worth noting that its growth has also been directly linked to policy supports, at least until recently. In this regard, the implementation of various support mechanisms, such as feed-in-tariffs, tax incentives or tradable green certificates, led to the remarkable growth of wind energy generation over the period in some European countries such as Germany and Spain, and later in others such as the USA, China and India [31].

Table 1. Growth rates and electricity consumption increase during 1990–2014. Source: Own production from IEA [25].

	Total energy growth rate percentage	Electricity growth rate percentage	Electricit increase (ktoe)	RE contribution to electricity increase percentage	WE contribution to electricity increase percentage	NRE contribution to electricity increase percentage
World	48.08	101.15	843384	18.65	5.22	76.13
EU15	6.34	32.07	50255	40.71	29.45	29.84
EU+	−21.37	6.85	2011	53.62	24.11	22.28
ET	−33.94	−14.96	−17239	10.81	0.18	89.00
ODC	15.61	37.53	132522	8.49	10.81	80.70
CA	102.12	199.33	4461	36.20	0.51	63.29
SA	60.10	54.30	7284	19.64	0.04	80.32
MENA	160.31	310.26	70187	5.39	1.06	93.55
EAS	177.29	668.51	437650	16.06	2.34	81.60
SAS	127.26	311.25	80576	10.38	2.70	86.92
LAC	81.43	150.30	61388	41.50	1.51	56.99
CAR	29.32	110.59	1922	4.75	0.44	94.81

The different growth rates by world regions makes the changes that took place evident, with regard to the weight of each of the sources used to generate the electricity used, in the period 1990–2014, especially for wind electricity. Firstly, the notable increase in the participation of the East Asian region in world electricity consumption, from 8% to 33%, may be highlighted. Likewise, South Asian and MENA regions also increased their participation significantly.

Secondly, only the East Asian region increased the renewable electricity world participation in a relevant manner, from 7% to 30%. The increases of renewable energy in electricity production in China may be related, according to Lo [32], to several factors, such as energy security (in order to reduce fossil fuel dependence), pollution and the increasing international pressure to control its carbon emissions, the growth of the manufacturing of renewable energy products (which are making China a global leader and exporter of green technologies) and finally, the consideration that renewable energies may be the appropriate way to bring electricity to underdeveloped rural areas.

Thirdly, the participation in wind electricity has changed remarkably. Thus, while in 1990, only three regions used wind electricity significantly (ODC: 80%, EU15: 19% and SAS: 1%), at the end of the studied period, many more regions use this energy. It is worth noting that the EU15 especially increased its participation, reaching 32%, which reflects the notable expansion of wind energy in the region, which, as mentioned before, may be related to wind energy support in the EU countries. Likewise, the East Asian region emerges strongly, representing 23% of total wind electricity used in the world, and other regions have started using wind electricity, although still with small percentages (EU+, MENA and LAC). The Chinese Renewable Energy Law, which came in force in 2006, prompted a large increase in renewable energy, particularly wind and photovoltaic power [33]. These renewable energy increases were made possible by renewable portfolio standards, feed-in tariff, and direct subsidies [32]. Also, it is worth noting that the growing participation of other world regions in worldwide wind energy may be related to their increasing demand for electrical energy, the continuous reduction of wind energy production costs, especially in recent years, which is placing it at competitive levels with respect to fossil fuels [34,35], and the allowance of new wind power plants through auctions involving companies with foreign capital and resulting in low bid prices [36].

Finally, it is also worth noting that only two regions increased their participation in non-renewable electricity consumption use, being the other developed countries and MENA regions which varied from 1% to 6%. These increases reflect that these regions have made less effort in the use of renewable energies to produce electricity. The OECD [37] reports indicate that there are three main barriers that slow investment in renewable energy in MENA countries. Firstly, the lack of profitability of renewable energy projects, which may also be related to the fact that many of these countries have fuel or gas energy resources. Secondly, insufficient cash flow to recover the high investment costs, and thirdly, difficulties in accessing financing. As stated in Kahia et al. [38], the MENA countries have high energy costs, which are related to MENA governments that typically sustain the energy market by subsidizing energy prices, creating barriers for new investors, thereby continuing to favor fossil fuels.

On the other hand, the other developed countries have also increased the non-renewable energies to produce electricity, which may be related not only to the fact that the USA did not commit itself to a reduction of carbon emissions or ratify the Kyoto Protocol [39], but also to the expansion of shale gas in its territory. As stated in Wang et al. [40], since 2000 the main oil and gas companies aggressively entered the shale gas business due to the increasing profitability of natural gas in shale formations production. Thus, lower natural gas prices have not only encouraged substitution away from sources such as coal and nuclear but also from renewable energies [41].

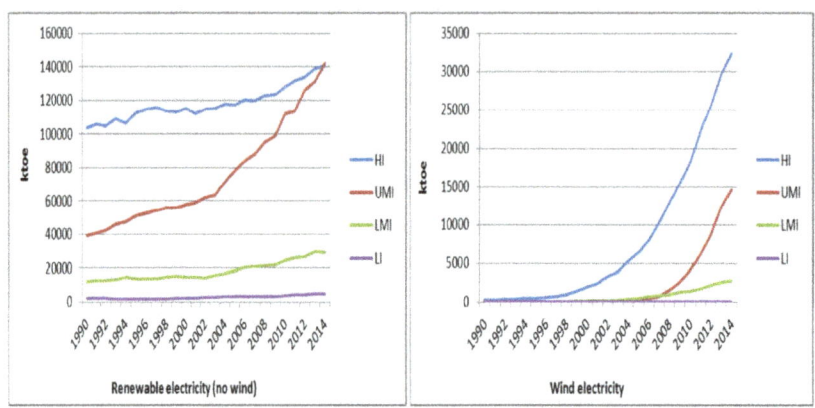

Figure 2. Renewable and wind electricity consumption trends (1990–2014). Economies by per capita GNI levels in 2014. Source: Own production from IEA [25].

Figure 2 shows the renewable and wind electricity consumption trends by per capita GNI levels. Renewable electricity consumption is only proved a practically stable consumption in lower income countries, also having a very low level. The rest of the regions presented a growing trend. This is especially notable for upper medium income countries, which reached the high income countries level in 2014. Wind electricity consumption showed an exponential rate of growth, as much as in high as in upper medium income countries, although in the latter countries this growth started later, around 2005, also reaching lesser values than the first. Likewise, lower medium income countries also grew, but in this case, the trend is linear. Additionally, its values are much smaller. Finally, lower income countries practically do not have wind electricity. Therefore, wind electricity use is growing especially in the more developed countries, probably due to the high investment costs of the technologies, which implies the need of some kind of state aid or subsides to promote it. It is worth noting that although the wind costs are declining as stated before, the project startup cost of a wind turbine is high [34].

In spite the growing trend in wind energy use over recent years, it is worth noting that it still represents a small percentage of total electricity consumption and therefore of total energy use. These values are represented in Figure 3, which shows the total energy, electricity and wind consumption in 2014 by world regions (colors) and by per capita GNI levels (plots). The area of the outer pie shows the total energy consumption in the region, while the area of the inner pie shows the electricity consumption, the latter being divided into non-renewable, renewable and wind electricity. Thus, by electricity sources, it can be clearly seen that most of the electricity has its origin in non-renewable energies. Only the Latin American and EU15 regions have more than a quarter of their electricity produced by renewable energies. With respect to wind electricity, only three regions (EU15, ODC and EAS) show appreciable electricity percentages, of which the EU15 has the highest. Additionally, Figure 3 shows that the ratio of electricity consumption to total energy decreases with a decrease in GNI. Likewise, wind electricity consumption decrease with a decrease in GNI.

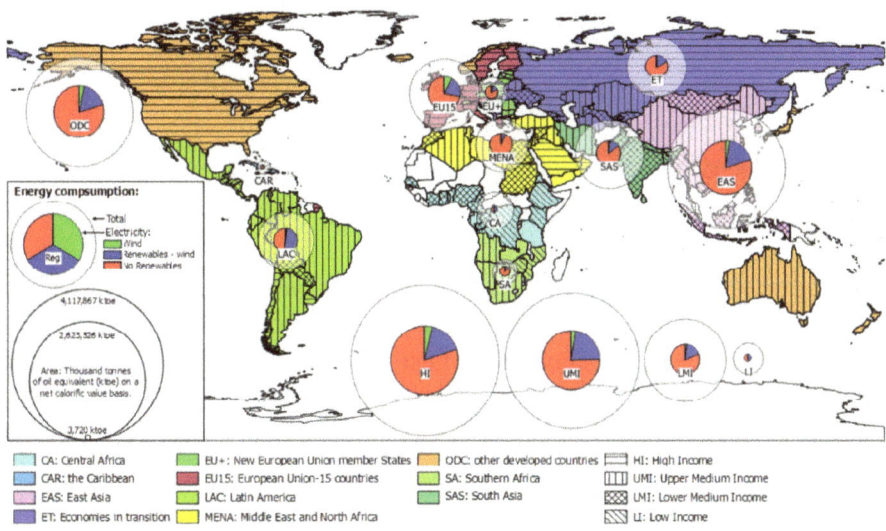

Figure 3. Total energy, electricity and wind electricity consumption in 2014. Source: Own production from IEA [25]. Maps © EuroGeographics for the administrative boundaries.

3.2. Total and wind electricity consumption per inhabitants during 1990–2014

Figure 4 shows the evolution of electricity consumption per inhabitants, by world regions in the period 1990–2014. During the period, the total electricity consumption per inhabitants was around

200 Ktoe/million inhabitants, showing a slightly positive trend. Above the world trend is highlighted the other developed countries region, with a 930 Ktoe/million inhabitants electricity consumption in 2014. During the period, its trend has been positive, although a negative trend since 2005 is observed. Likewise, EU15 also shows values notably higher than the world, with a trend similar to the other developed countries region. The economies in transition and new EU members regions can also be seen to be above the world trend. Both began the period with decreasing electricity consumption, but started growing from 1995 with the world trend. Below world trend, it is worth mentioning the high growth in the East Asian region, which surpassed the world value at the end of the period. Likewise, MENA, Latin American and South Asian regions also show positive trends.

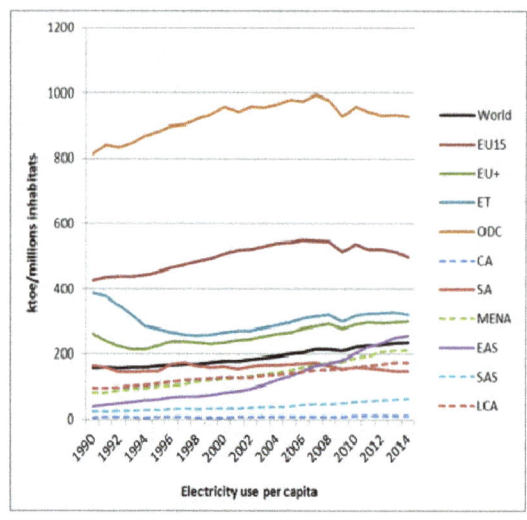

Figure 4. Total electricity per capita consumption (1990–2014): World Regions. Source: Own production from IEA [25].

It is again worth noting that urbanization and economic development are strictly linked to the variations in electricity consumption [42]. Urbanization and economic growth may be two of the main causes of the electricity consumption in East Asian countries, while the declining trend observed in other developed countries and the EU region since 2008, may be related to the decrease of economic growth linked to the financial crisis. Nevertheless, in spite of this electricity consumption decrease, it is observed that other developed countries and the EU countries have higher electricity per capita consumption, which is in line with the conclusions of Ferguson et al. [28]. These authors state that, for the global economy as a whole, there is a stronger correlation between electricity use and wealth creation for richer countries, therefore implying that these countries will have higher electricity consumption in per capita terms.

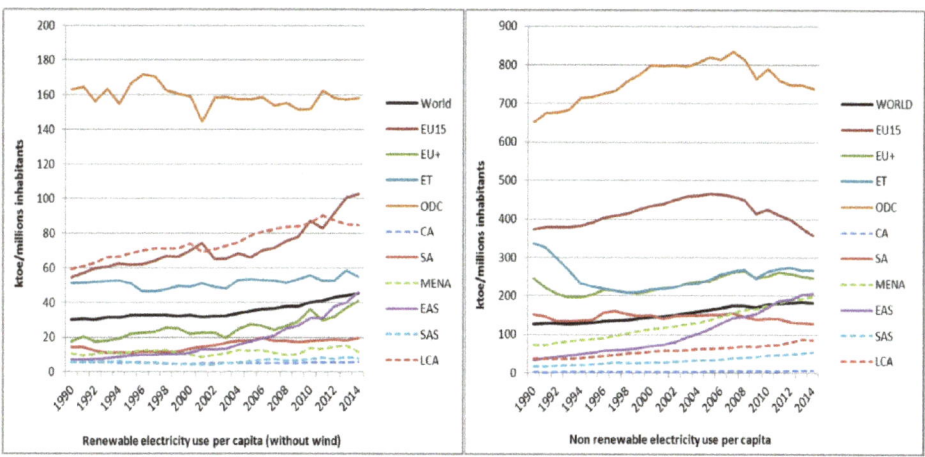

Figure 5. Renewable and non-renewable electricity per capita consumption trends (1990–2014): World Regions. Source: Own production from IEA [25].

Figure 5 shows the evolution of renewable and non-renewable electricity consumption per inhabitants, by world regions along the studied period. As can be observed, the renewable electricity consumption trend is positive globally, although, it did not start growing until 2002. Once again, below the world trend, the other developed countries region is highlighted, which shows a quite constant trend during the period. Additionally, it may be pointed out the growing trend of EU15 and Latin American regions (below the world trend). Likewise, the East Asian and new EU members regions also have a positive trend. Although their values are smaller than world figures, at the end of the period both regions presented similar values to the world region. Nevertheless, these increasing trends may have different explanations. The increasing trend observed for all EU countries may be explained by the European energy and climate policies aimed at promoting renewable energies. As stated before, the EU has been playing a leading role in setting ambitious targets for renewable energy and establishing appropriate policies and instruments to support it, especially oriented to electricity production. Therefore, the decreasing trend in non-renewable energy consumption is partially due to its substitution by renewable energies. These changes in the electrical energy mix are proved to be the main factor driving down CO_2 emissions in the EU countries, at least since 2000 [43]. However, in the East Asian region, the renewable energy increase may be related more to increased electricity demand than to changes in the energy mix. Both graphs in Figure 5 show increasing trends for Asian regions, implying that all fuels have been growing to satisfy the electricity demand. In this regard, it is worth noting that renewable energy, especially solar and wind energy, can be installed in rural regions where it is difficult to bring electricity. Additionally, some renewable plants can be in operation in a few months instead of some years [35]. Finally, the renewable energy growth in the Latin American region is also due to economic growth and urbanization and the rise of the middle class, but is especially linked to hydropower electricity [44]. The study by Al-Mulali et al. [45] indicated that renewable electricity consumption is more significant than non-renewable electricity consumption in promoting economic growth in the Latin American region.

On the other hand, it is worth noting that the regional trends for non-renewable electricity use are quite similar to those for electricity, was showed in Figure 4. Nevertheless, it may be pointed out that the MENA region ended the period with more non-renewable electricity consumption per capita than the world region. The MENA countries are using non-renewable energies to produce electricity in order to satisfy their increasing demand, which are mainly related to urbanization and economic growth. Thus, in spite of having high solar irradiation and strong winds, offering excellent potential for generating electricity from renewable sources, the use of renewable energy for generating electricity still remains low [46].

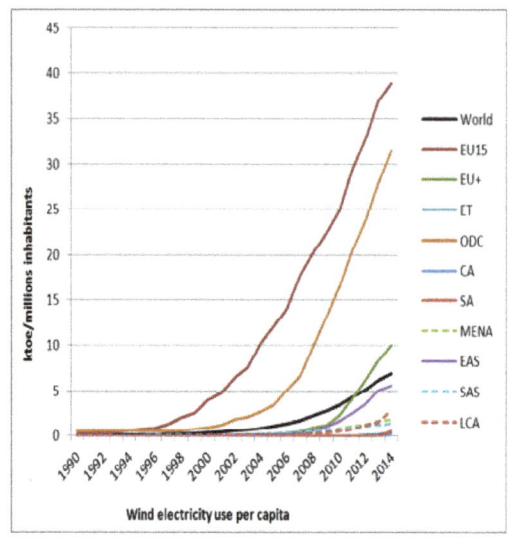

Figure 6. Wind electricity per capita consumption trends (1990–2014): World Regions. Source: Own production from IEA [25].

Finally, Figure 6 shows a very different behavior with respect to wind electricity use. At the beginning of the period, hardly any region used wind electricity to produce electricity. However, some of them started producing this energy during the period. The first region, which started increasing wind electricity consumption, was the EU15, around 1995. From this year, the region showed exponential growth. The second region increasing its wind electricity use was other developed countries, at the end of the nineties (also with an exponential growth). Its growth was very similar to that of the EU15, but always with lower values. Later, around 2006, two regions also started increasing their wind electricity production with an exponential growth: the new EU members and East Asian regions. Nevertheless, the first reached higher growth. Additionally, Latin American and South Asian countries have also increased their wind electricity consumption since 2006, but with a linear trend. Finally, the strongest growth in wind energy in the MENA region occurred between 2005 and 2010. Nevertheless, the socio-political events linked to the Arab Spring in some parts of the MENA region have slowed the incipient development of wind energy [46]. Therefore, it is worth noting the leading role of the EU during the studied period. Nevertheless, Figure 6 also shows the rise of the other developed countries and the growth of the wind energy industry in the East Asian region, especially since 2005. In addition, the gradual adoption of wind energy by several countries of the developing world is of special interest, and demonstrates the ability to largely substitute fossil-fueled power generation [31]. Along this line, wind energy has achieved remarkable advances and nowadays most studies show that it is an economically efficient and accessible energy resource for large scale utilization because of the decreasing energy cost [34]. Thus, according to the REN21 [36] report, the weighted average investment cost for onshore wind fell by slightly more than two-thirds between 1983 and 2015. Therefore, this decreasing cost, the growing interest of developing countries in promoting cleaner energy sources and greater energy security to meet the growing demand for energy, are factors that may be behind the growing development of wind energy in the developing regions of the world.

Notable electricity consumption differences are also shown when countries are grouped by per capita GNI levels. Figure 7 shows that the highest electricity per capita consumption is conducted, with an outstanding difference, by high income countries. Likewise, it is observed that the electricity consumption in per capita terms is lower as income decreases. Therefore, as observed by Kim [47], groups with relatively high per capita income tend to exhibit high electricity consumption. In addition, the lack of overall convergence in per capita income is observed to be an important source of per capita electricity consumption divergence.

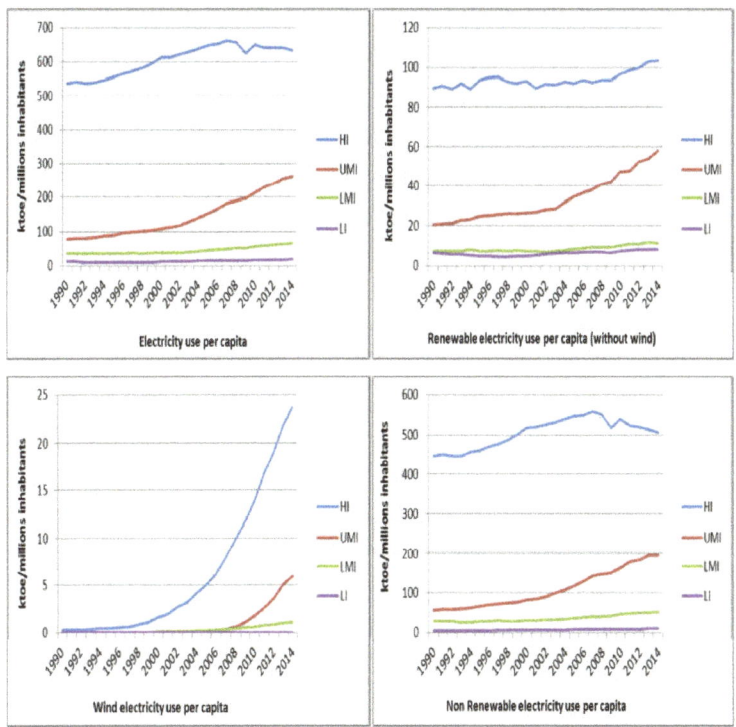

Figure 7. Total and residential energy per capita consumption trends (1993–2013). Economies by per capita GNI levels in 2014. Source: Own production from IEA [25].

Figure 7 also shows that the high income group electricity consumption trend is positive until 2006 and then slightly decreasing. These countries also show a similar trend for non-renewable electricity consumption, being constant until 2008 and positive since then for renewable one. The total electricity consumption decreasing trend may be explained not only by GDP decline, but also by energy efficiency gains. It is worth noting that the electricity consumption decrease is linked to non-renewable energies, as renewable energies are still having a positive trend, even after 2006. Figure 7 shows that upper middle-income countries have a positive trend for non-renewable and renewable electricity consumption per capita. In this regard, Apergis and Payne [26] found that the factor that affects electricity consumption in these countries is mainly urbanization. It is also relevant to note that the lowest income countries have low electricity consumption in per capita terms, without growth. In these countries, as stated by Apergis and Payne [26], the variations in electricity consumption may have an impact on both GDP and urbanization levels, with the electricity shortages which make these countries very vulnerable being one of the most relevant problems observed.

Finally, it is worth noting the special growth of wind electricity consumption presented by the groups. Once again, the high income countries stand out in the graph, but this time with a clear exponential trend. Later, around 2006, the upper middle income countries also started growing with an exponential trend. The lower middle income countries also started growing from 2006, but with a linear trend. Finally, no growth is observed for the lowest income countries.

3.3. Total and wind electricity consumption per GDP during 1990–2014

Figure 8 shows the evolution of electricity and renewable electricity consumption in terms of GDP, by world regions in the period 1990–2014. The world trend is quite constant for the electricity and renewable energy consumption per GDP. Nevertheless, those regions with higher electricity consumption per GDP (ET and SA) had notable decreasing trends since 1995. In this regard, Brizga et al. [48] find a significant correlation between the reindustrialization process and the energy intensity decrease in the economies in transition. Likewise, the industrial efficiency gains may be related to the South African countries, as in the first stage of development the industrial improvement may not be compensated by household energy demand growth [49]. On the other hand, the East Asian region presents a positive trend, its values being the highest at the end of the period. The rapid industrialization of this region, especially in energy intensive industries, and the rapid household electricity demand growth, due to notable urbanization and changes in lifestyle, may be the cause of the growing trend. Finally, it is also worth noting the decreasing trend for the developing countries group. The electricity intensity decrease may be related not only to energy efficiency policies, but also to an economy dematerialization and to industry displacement to developing countries [49,50].

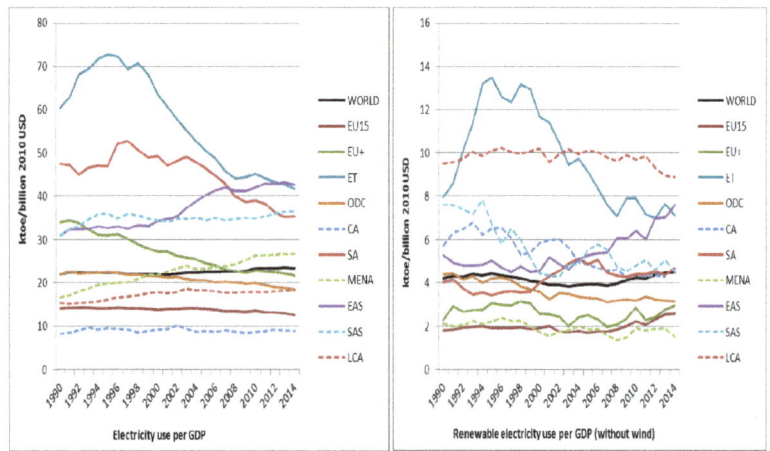

Figure 8. Total and renewable electricity use per GDP (1990–2014): World Regions. Source: Own production from IEA [25].

In terms of renewable energies, transition and East Asian countries show a similar trend, which may be explained in the same way. However, South African countries show a quite constant trend during the whole period, which implies no main changes in the energy mix. More remarkable is the trend for wind electricity showed in Figure 9. An exponential growth is observed in almost all regions. Above the world trend, are highlighted the EU15 and the Asian countries groups, with the East Asian group showing later, but more intense, growth. In this regard, it is worth noting the rapid incorporation of wind energy into the economy.

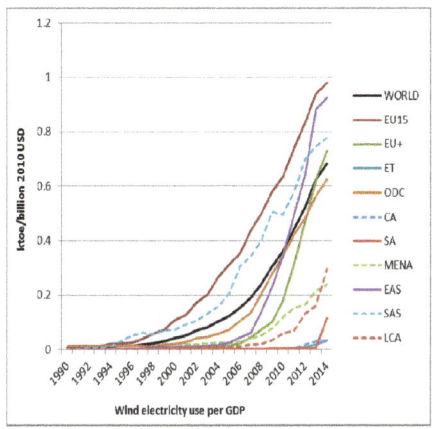

Figure 9. Wind electricity use per GDP (1990–2014): World Regions. Source: Own production from IEA [25].

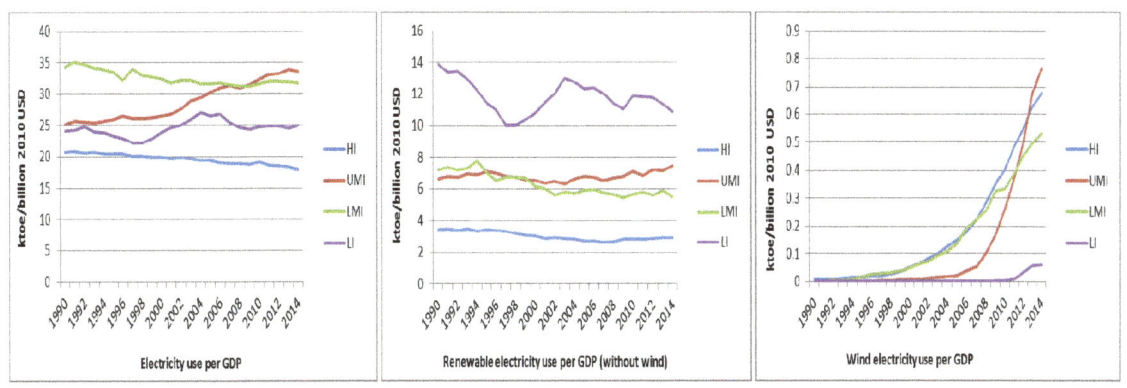

Figure 10. Total, residential and wind energy use by GDP consumption trends (1993–2014): Economies by per capita GNI levels in 2014.

Finally, Figure 10 shows that electricity and renewable electricity consumption per GDP tend to be decreasing in all regions when considering the economies of the groups by per capita GNI, except for upper medium income region which presents a remarkable increasing trend for electricity and a slightly increasing trend for renewable electricity, which may be related to the increasing electricity consumption by households. Again, it is worth noting the exponential growth of wind energy. The upper medium income region shows a later and more intense growth, while the lower income countries only start to grow slightly at the end of the period. Additionally, it is worth noting that while renewable energy with respect to GDP is lower for high income countries, the wind energy is almost higher for them through the whole period, which may be related to the high biomass use in developing countries. Therefore, it may be highlighted that over recent years, renewable energy has become more and more dominated by wind and solar [35].

3.4. Electricity and wind consumption ratios during 1990–2014

Figure 11 shows the ratio of electricity to total energy consumption and total renewable electricity to total electricity consumption by the world. World electricity consumption represents around 18% of total energy consumption at the end of the period. During the period, the world trend of the electricity to total energy consumption ratio has been positive. All regions show a positive trend, except Southern African countries which shows a decreasing trend from 2008. It is worth noting the high growth of this ratio, in the case of the East Asian region, whose rate grew from 5% to 20%. It is also worth noting that the EU15 and other developed countries regions show the highest rate value (around 23%), while the Central African shows 3%.

The ratio of renewable electricity to total electricity consumption grew slightly in the world from 2002, being around 23% in 2014. Most regions also showed a similar growth trend from 2002, although three regions showed negative trends. On the one hand, Latin American and Central African regions showed a notable negative growth, but still had very high ratio values at the end of the period, of around 50%. On the other hand, the MENA countries presented a softer negative trend but with only an 8% ratio value at the end the period, being the lowest. In addition, it is worth noting that the Southern African region had a more pronounced positive trend.

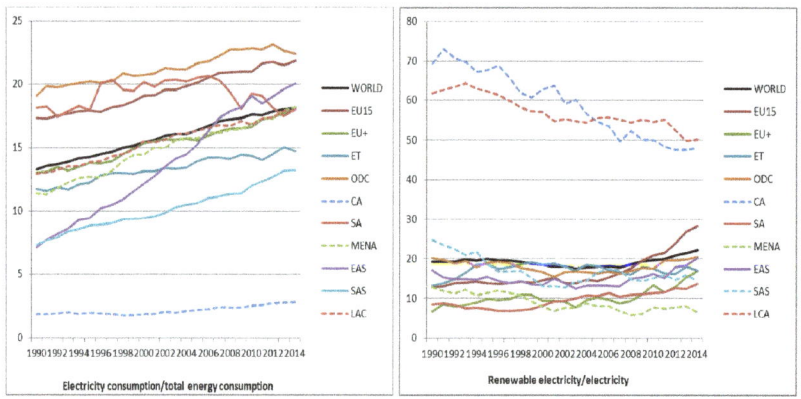

Figure 11. Total and renewable electricity ratios by world regions.

Figure 12 shows the ratio of wind electricity to renewable electricity consumption. An exponential growth is shown in the world, starting from almost 0% to 13%. The EU15 is the region with the highest growth, ending the period with 28%. A remarkable growth is also observed in the new European countries region, especially from 2010 to 2012. Asian, MENA and other developed countries had similar trends, although East Asian and other developed countries started growing earlier. At the end of the period, their ratio ranged from between 12% to 16%. The other regions only showed a slight growth in the latter years.

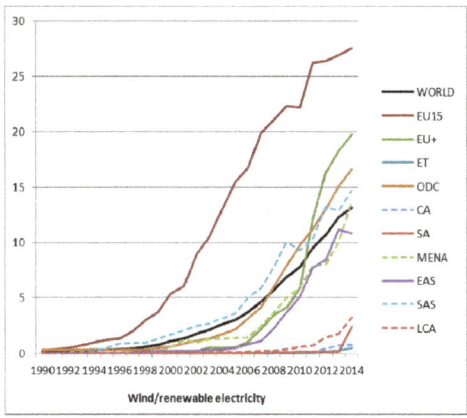

Figure 12. Wind ratio by world regions.

Figure 13 shows notable differences in these ratios when considering the world regions by per capita GNI level. Electricity consumption with respect to total energy is around 21%, 19%, 14% and 6% in high, upper medium, lower medium and lower income countries, respectively, in 2014, increasing in all cases during the period. The renewable electricity with respect to electricity ratio is quite constant in lower medium income countries, slightly increasing in high income countries from 2005 and slightly decreasing in upper medium income countries, but all of them ended the period with a ratio ranging from 20% to 22%. Lower income countries presented a higher ratio in 2014 (45%). Finally, wind electricity to renewable electricity consumption had an exponential growth in high and upper medium income countries, the second starting to grow around 2005. Lower medium income countries show a constant linear growth and lower income countries have only a very slight growth from 2012. In 2014, the ratio values are equal to 19% in high income countries, 9% in upper medium and lower medium income countries and less than 1% in lower income ones. It is worth noting that the lower percentage of wind with respect to other renewable energies in non-high-income countries, at the end of the period, is due to the relevance of biomass and hydropower. Nevertheless, once again it is observed that wind energy, along with solar, is becoming one of the most important renewable energy sources to generate electricity [35].

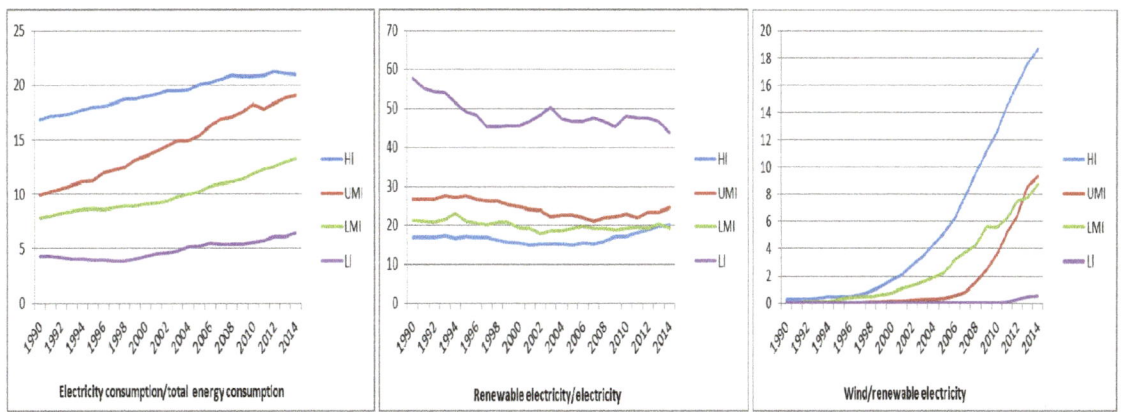

Figure 13. Renewable electricity and wind ratios by per capita GNI level in 2014.

4.   Wind Electricity Trends Explaining Factors

Economic growth is one of the main causes of increased energy use and especially of electricity consumption increases over recent times, as has been noted in several previous papers, such as in Apergis and Payne [26]. Nevertheless, urbanization, changes in lifestyles and improvements in electricity access are also relevant factors, especially for developing countries. Thus, the economic growth and these other factors have been increasing the electricity consumption worldwide, being particularly relevant in those regions in which these changes have happened in a more intensive way. The increasing electricity demand has provoked the growth of electricity production, both from non-renewable and renewable energies.

Nevertheless, these changes have also been accompanied by a growing interest in climate change and environmental issues, which has led governments to have more interest in the use of renewable energy as a source of energy generation in general, and electricity in particular. At first, attention was focused on the most developed countries, as they were considered the main causes of pollution and climate change. Thus, the Kyoto Protocol [51] set binding emission reduction targets only for 37 industrialized countries and the European community over the five-year period, 2008 to 2012, which led those countries to promote renewable energies and energy efficiency even before the Kyoto protocol entered into force. Hence, during the period 2005–2012, Europe recorded the highest share in total new global investments in renewable energies, with wind, solar and biomass industries being especially supported [52]. Developing countries, including China, India, Brazil and South Africa, did not have to reduce their emissions, but were encouraged to also adopt policies to promote renewable energies. Thus, for example, China approved the Chinese Renewable Energy Law in 2006, mainly boosting wind and photovoltaic energy [33]. Over time, the role of the developing countries in the fight against climate change and in the promotion of renewable energies has been increasing, which can be highlighted with the signing of the Paris agreement at COP21. Thus, many countries worldwide have begun to develop energy plans to combat climate change, which has given a significant boost to the growth of renewable energy, especially solar and wind [35].

However, the exponential growth of wind energy cannot be explained solely by increasing concern about climate change worldwide. Firstly, it is worth noting that wind energy has grown notably in those countries with high or increasing electricity consumption and without their own fuel or gas resources. Security motivations and energy independence gains have also motivated the renewable energy investments. Thus, energy security is especially relevant for those countries where growing dependence on imported gas is a significant energy security issue. In this regard, renewable energy can provide alternative sources of electric power [53]. It may also be an appropriate way to bring electricity to rural areas which have difficulty integrating into the distribution network [32]. Nevertheless, it is worth noting that some renewable energy technologies, such as wind, depend on different natural cycles, and are therefore subject to variability on differing timescales. Thus, it is important to have alternative electricity plants, and to encourage inter-connections between grids and across national borders to balance supply and demand. Some interconnectors have been built, such as Skagerrak 4, which helps to balance Denmark's wind and thermal power and Norway's hydropower. Likewise, innovative hybrid systems linking hydropower and renewable plants have also emerged [54]. Moreover, stationary battery storage continues to advance, and costs are decreasing, although most of the capacity has been installed in the developed world. However some storage projects are also being undertaken in developing countries, particularly in conjunction with mini-grids [36].

Additionally, wind deployment was also driven by wind power's cost-competitiveness. The continuously decreasing cost has determined that wind energy is becoming the least-cost option for new power generating capacity in an increasing number of markets [36]. However, as stated by Kumar et al. [34], wind energy systems have very high start-up costs, needing a large scale investment. Therefore, support measures have been adopted by governments to promote wind energy, usually being performed by the combination of several measures such as Feed-in-Tariff (FIT), Fit-in-Premiums (FIP), tenders, quota obligations or contracts for difference. FIT systems are the most used worldwide because of their better dynamic efficiency, low transaction cost, diversity, and high investing securities. Nevertheless, it may not be able to respond quickly to falls in production costs, generating overcompensation and excessive demand for new installations [55]. Therefore, authors such as Hiroux and Saguan [56] consider that wind energy producers should be exposed to market signals, FIP therefore being a more appropriate choice. Alternatively, auctions are emerging as a popular alternative because of their potential to achieve deployment in a cost-efficient and regulated manner, nevertheless they have to be adequately defined [57]. Also, the degree of effectiveness of these policies may vary among countries and regions, even though establishing a stable legal framework can be considered the best way to promote investments in renewable energies. Issues related to the political and economic framework are the most relevant barriers for wind energy diffusion [58].

Finally, it is worth noting that support measures may be difficult to implement in the poorest countries due to their low budgets. International aid is needed in order to establish adequate promotion. Alternatively, these countries may also consider the opportunity to promote foreign investments to enhance renewable energy projects, which could improve the electricity supply in their country, boosting their economic development

5. Conclusions and Policy Implications

Wind energy is crucial to achieve CO_2 emission reductions related to electricity production as it is believed to have the least adverse environmental impact and is one of the renewable energies that has become economically affordable long before several others. Nevertheless, although wind power has begun to grow, especially from 2005, it still represents a small share of electricity consumption.

At regional level, remarkable differences are observed between the world regions. Therefore, diverse global energy policy should be applied. All regions, except Southern African, have electricity growth rates higher than that of total energy in the analyzed period, the regions with the highest growth rates being in Asian and MENA regions. The electricity growth was mostly produced by non-renewable energies, with 18.65% generated by using RE and 5.2% by using wind energy, with the highest renewable electricity consumption participation being observed in new European Union members, Latin American and EU15 countries, and the highest participation of wind electricity in EU15, new European Union members, and the other developed countries. Very little wind electricity was used in Southern African, economies in transition, Central African and The Caribbean regions. In these regions, this participation is less than 1%. Therefore, all regions should make greater effort if they want to increase renewable energies in order to reduce emissions, especially those with the highest electricity consumption and not much renewable use, for example, the South Asian and

MENA regions. It is convenient to make efforts to change the electricity mix by discouraging fossil fuel use and promoting renewable and also wind energy use. The EU regions are the only ones with a notable use of renewable energies, and wind, which may be related to their current energy policy, nevertheless, it is also desirable to take a closer look at their energy policies.

The different evolution in the electricity consumption and in the main generation sources between 1990 and 2014 have changed their distribution in the world. In electricity consumption, it is worth noting the notable participation increase in East Asian region, and to a lesser value, in South Asian and MENA countries. However, renewable electricity consumption also increased in East Asian countries, but not in the other regions, which means they need to make special efforts to promote their renewable energies for electricity generation. On the other hand, the participation in wind electricity has changed remarkably while, as in 1990, only three regions used wind electricity (ODC, EU15 and SAS), in 2014, East Asian, new EU members, MENA and Latin American countries also used it. Likewise, it is worth noting that a notable expansion of wind electricity use is observed in EU15 and also in East Asia (23% of total wind electricity). Therefore, countries should focus on the behavior of these regions if they want to impulse their wind power usage.

In per capita terms, the renewable electricity consumption trend is positive in the world, starting from 2002. EU15, Latin American, East Asian and new EU members countries have positive trends, while MENA countries end the period with more non-renewable electricity use per capita than the world region. Referring to wind electricity, at the beginning of the period, hardly any region used wind electricity to produce electricity. The first regions which started increasing it were EU15 and other developed countries which showed exponential growth. Later, around 2006, new EU members and East Asian region also started to grow exponentially. Finally, Latina American and South Asian countries have also increased their wind electricity consumption from 2006, but with a linear trend. By per capita GNI levels, high income countries stand out in the use of wind electricity, with a clear exponential trend. Later, upper medium income countries also started growing, also with an exponential trend, while lower medium income countries had a linear trend. Finally, no growth is observed for lower income countries. More effort is especially needed in lower medium and lower income countries. Nevertheless, as the exponential growth observed in more developed countries has been promoted with important support measures, international aid will be required in the less developed countries in order to increase renewable and wind electricity to produce electricity.

In GDP terms, the world trend is quite constant for electricity and renewable electricity consumption per GDP. Nevertheless, those regions with higher electricity consumption per GDP (ET and SA) have had notable decreasing trends since 1995, while East Asian region presents a positive trend, its values being the highest at the end of the period. The trend for wind electricity is exponential in almost all regions, highlighting EU15 and Asian regions. By per capita GNI groups, electricity and RE per GDP tend to be decreasing in all regions except for the upper medium income countries which present a remarkable increasing trend for electricity and a slightly increasing trend for renewable electricity use.

Acknowledgments

The first author acknowledges the funding received from the Project SEJ 132 project by the Andalusian Regional Government, the funding received from the Project ECO2014-56399-R by

Spain's Ministry of Economy and Competitiveness, from the "Cátedra de Economía de la Energía y del Medio Ambiente" (Department for Energy Economics and the Environment) at the University of Seville (Reference: 1394/0103) and the "Fundación Roger Torné", and the funding received from the Project FONDECYT Regular 1150025 of Chile's Ministry of Education. The standard disclaimer applies.

Conflict of Interest

All authors declare no conflicts of interest in this paper.

References

1. Burleson E (2016) Paris agreement and consensus to address climate challenge. Social Science Electronic Publishing.
2. UNFCCC (2015) Adoption of the Paris agreement. Available from: https://unfccc.int/resource/docs/2015/cop21/eng/l09r01.pdf.
3. Climate Action Tracker (2015) Paris agreement: near-term actions do not match long term purpose-but stage is set to ramp up climate action. Available from: http://climateactiontracker.org/assets/publications/briefing_papers/CAT_COP21_Paris_Agreement_statement.pdf.
4. Brauers H, Richter PM (2016) The Paris climate agreement: is it sufficient to limit climate change? DIW Roundup: Politik im Fokus, No. 91.
5. IEA, CO_2 emissions from fuel combustion. International Energy Agency, 2015. Available from: https://www.iea.org/publications/freepublications/publication/CO2EmissionsFromFuelCombustionHighlights2015.pdf.
6. IEA, Energy and climate change. International Energy Agency, 2105. Available from: https://www.iea.org/publications/freepublications/publication/WEO2015SpecialReportonEnergyandClimateChange.pdf.
7. Premalatha M, Abbasi T, Abbasi SA (2014) Wind energy: increasing deployment, rising environmental concerns. *Renew Sust Energ Rev* 31: 270–288.
8. IEA (2016) Electricity Information 2016. Paris: IEA.
9. Scarlat N, Dallemand JF, Monforti FF, et al. (2015) Renewable energy policy framework and bioenergy contribution in the European Union-an overview from national renewable energy action plans and progress reports. *Renew Sust Energ Rev* 51: 969–985.
10. Dent CM (2013) Wind energy development in East Asia and Europe. *Asia Eur J* 11: 211–230.
11. Mohammed YS, Mustafa MW, Bashir N (2013) Status of renewable energy consumption and developmental challenges in Sub-Sahara Africa. *Renew Sust Energ Rev* 27: 453–463.
12. Chingulpitak S, Wongwises S (2014) Critical review of the current status of wind energy in Thailand. *Renew Sust Energ Rev* 31: 312–318.
13. Ho LW (2016) Wind energy in Malaysia: past, present and future. *Renew Sust Energ Rev* 53: 279–295.
14. Kaplan YA (2015) Overview of wind energy in the world and assessment of current wind energy policies in Turkey. *Renew Sust Energ Rev* 43: 562–568.

15. Leung DY, Yang Y (2012) Wind energy development and its environmental impact: a review. *Renew Sust Energ Rev* 16: 1031–1039.

16. Mundaca L, Markandya A, Nørgaard J (2013) Walking away from a low-carbon economy? Recent and historical trends using a regional decomposition analysis. *Energ Policy* 61: 1471–1480.

17. IEA, Key electricity trends. International Energy Agency, 2016. Available from: https://www.iea.org/publications/freepublications/publication/KeyElectricityTrends.pdf.

18. IEA (2016) Renewables Information 2016. Paris: IEA.

19. IEA, Key renewables trends. International Energy Agency, 2016. Available from: https://www.iea.org/publications/freepublications/publication/KeyRenewablesTrends.pdf.

20. Bridge G, Bouzarovski S, Bradshaw M, et al. (2013). Geographies of energy transition: space, place and the low-carbon economy. *Energ Policy* 53: 331–340.

21. Pablo RMP, Pozo BR, Yñiguez R (2017) Global changes in residential energy consumption. *Energ Policy* 101: 342–352.

22. World Bank (2017). World Bank Open Data. Available from: http://data.worldbank.org/.

23. United Nations (2015) World Economic Situation and Prospects 2015. New York: United Nations. Available from: http://www.un.org/en/development/desa/policy/wesp/wesp_ archive/2015wesp_ full_en.pdf.

24. IEA, Monthly electricity statistics. International Energy Agency, 2016. Available from: https://www.iea.org/statistics/monthlystatistics/monthlyelectricitystatistics/.

25. IEA, Statistics. International Energy Agency, 2016. Available from: http://www.iea.org/statistics/.

26. Apergis N, Payne JE (2011) A dynamic panel study of economic development and the electricity consumption-growth nexus. *Energ Econ* 33: 770–781.

27. Ferguson R, Hill R, Craggs C, et al. (1997) Benefits of electricity generation. *Eng Sci Educ J* 6: 255–259.

28. Ferguson R, Wilkinson W, Hill R (2000) Electricity use and economic development. *Energ policy* 28: 923–934.

29. European Commission (1997) Communication from the Commission. Energy for the future: renewable sources of energy. Available from: http://europa.eu/documents/comm/white_ papers/pdf/com97_599_en.pdf.

30. Kitzing L, Mitchell C, Morthorst PE (2012) Renewable energy policies in Europe: converging or diverging? *Energ Policy* 51: 192–201.

31. Kaldellis JK, Zafirakis D (2011) The wind energy (r) evolution: a short review of a long history. *Renew Energ* 36: 1887–1901.

32. Lo K (2014) A critical review of China's rapidly developing renewable energy and energy efficiency policies. *Renew Sust Energ Rev* 29: 508–516.

33. Schuman S, Lin A (2012) China's renewable energy law and its impact on renewable power in China: progress, challenges and recommendations for improving implementation. *Energ policy* 51: 89–109.

34. Kumar Y, Ringenberg J, Depuru SS, et al. (2016) Wind energy: trends and enabling technologies. *Renew Sust Energ Rev* 53: 209–224.

35. UNEP (2016) Global trends in renewable energy investment 2016. Frankfurt: United Nations Environment Programme-Frankfurt School of Finance & Management gGmbH 2016.

36. REN21 (2016) Renewables 2016 global status report. Paris: REN21. Available from: http://www.buildup.eu/en/practices/publications/renewables-2016-global-status-report.

37. OECD (2013) Renewable energies in the Middle East and North Africa: policies to support private investment, OECD Publishing.

38. Kahia M, Kadria M, Aissa MSB, et al. (2017) Modelling the treatment effect of renewable energy policies on economic growth: evaluation from mena countries. *J Cleaner Prod*, In press.

39. Biermann F, Brohm R (2004) Implementing the Kyoto protocol without the USA: the strategic role of energy tax adjustments at the border. *Clim Policy* 4: 289–302.

40. Wang Q, Chen X, Jha AN, et al. (2014) Natural gas from shale formation—the evolution, evidences and challenges of shale gas revolution in United States. *Renew Sust Energ Rev* 30: 1–28.

41. Newell RG, Raimi D (2014) Implications of shale gas development for climate change. *Environ Sci Technol* 48: 8360–8368.

42. Karanfil FLY (2015) Electricity consumption and economic growth: exploring panel-specific differences. *Energ Policy* 82: 264–277.

43. Karmellos M, Kopidou D, Diakoulaki D (2016). A decomposition analysis of the driving factors of CO_2 (Carbon dioxide) emissions from the power sector in the European Union countries. *Energy* 94: 680–692.

44. Balza L, Espinasa R, Serebrisky T (2016) Lights on: energy needs in latin america and the caribbean to 2040. Inter-American Development Bank.

45. Al-Mulali U, Fereidouni HG, Lee JY (2014) Electricity consumption from renewable and non-renewable sources and economic growth: evidence from Latin American countries. *Renew Sust Energ Rev* 30: 290–298.

46. REN21 (2013). MENA Renewables Status Report, Paris: REN21.

47. Kim YS (2015). Electricity consumption and economic development: are countries converging to a common trend? *Energ Econ* 49: 192–202.

48. Brizga J, Feng K, Hubacek K (2013) Drivers of CO_2 emissions in the former Soviet Union: a country level IPAT analysis from 1990 to 2010. *Energy* 59: 743–753.

49. Chang Y, Choi Y, Kim CS et al. (2016) Disentangling temporal patterns in elasticities: a functional coefficient panel analysis of electricity demand. *Energ Econ* 60: 232–243.

50. Pablo RMP, Sánchez BA (2017) The changing of the relationships between carbon footprints and final demand: panel data evidence for 40 major countries. *Energ Econ* 61: 8–20.

51. United Nations (1998). Kyoto protocol to the United Nations framework convention on climate change. *Rev Eur Commun Int Environ Law* 7: 214–217.

52. EEA (2016). Renewable energy in Europe 2016 recent growth and knock-on effects. Luxemburg, European Environment Agency. Available from: http://admin.indiaenvironmentportal.org.in/files/file/Renewable%20energy%20in%20Europe.pdf.

53. Ölz S, Sims R, Kirchner N (2007). Contribution of renewables to energy security. Available from: http://www.qualenergia.it/UserFiles/Files/contribution%20of%20renewables%20to%20energy%20security.pdf.

54. IHA (2016) 2016 Key Trends in Hydropower. London: International Hydropower Association.

55. Serrano GJ, Lacal AR. (2016). Technological evolution of onshore wind turbines-a market based analysis. *Wind Energ* 19: 2171–2187.

56. Hiroux C, Saguan M (2010). Large-scale wind power in European electricity markets: time for revisiting support schemes and market designs? *Energ Policy* 38: 3135–3145.

57. Lucas H, Ferroukhi R, Hawila D (2013). Renewable energy auctions in developing countries, International Renewable Energy Agency (IRENA).

58. Boie I, Held A, Ragwitz M (2014). Barriers and drivers framing the diffusion of renewable energy technologies. In: renewables in the EU: policy performance, drivers and barriers. 1st DIACORE-CEPS Policy Workshop.

10

Optimization of regular offshore wind-power plants using a non-discrete evolutionary algorithm

Angel G. Gonzalez-Rodriguez [1,*], Manuel Burgos Payan [2], Jesús Riquelme Santos [2], and Javier Serrano Gonzalez [2]

[1] Department of Electronic Engineering and Automation, University of Jaen, Jaen, Spain
[2] Department of Electrical Engineering, University of Seville, Camino de los Descubrimientos s/n, Seville, Spain

* **Correspondence:** Email: agaspar@ujaen.es

Abstract: Offshore wind farms (OWFs) often present a regular configuration mainly due to aesthetical considerations. This paper presents a new evolutionary algorithm that optimizes the location, configuration and orientation of a rhomboid-shape OWF. Existing optimization algorithms were based on dividing the available space into a mess of cells and forcing the turbines to be located in the centre of a cell. However, the presented algorithm searches for the optimum within a continuous range of the eight parameters that define the OWF, which allows including a gradient-based local search operator to improve the optimization process. The study starts from a review of the economic data available in the bibliography relative to the most significant issues influencing the profitability of the investment in terms of the Internal Rate of Return (IRR). In order to address the distinctive characteristics of OWFs, specific issues arise which have been solved. The most important ones are: interpretation of nautical charts, utilization of the seabed map with different load-bearing capacities, and location of the shoreline transition.

Keywords: wind energy; offshore; non-discrete evolutionary algorithm; continuous evolutionary algorithm; IRR; regular patterns; optimal configuration; gradient-based local search

1. Introduction

One of the keys to success in the development of (onshore) wind energy technology was the chance to test small parks, or even isolated wind turbines. Naturally, it is true that economies of scale apply to onshore wind parks to the extent that, in regions where large suitable and available surfaces exist, only big parks have been installed. However, in marine sites, the wide availability of surface area and the high fixed costs involved, lead to intensive resource exploitation, resulting in the installation of a large concentration of wind turbines. This enforces the application of economies of scale in order to capitalize on this type of facilities, in which the companies involved usually protect their knowledge in the form of confidential data. The sluggish development of offshore wind power plants is partly due to the fact that they are still largely seen as a risky investment and are therefore employed only in countries where most of the available onshore wind resources are already being used.

In order to reduce the risk, once the area that houses the wind farm has been selected, and before undertaking a thorough study of investment feasibility, a search tool is necessary to optimize the profitability of the park over its lifespan, which is a compromise between minimizing the initial investment and maximizing the annual revenue from energy sales. These values, initial investment and annual revenue are determined by the configuration of the park: a suitable choice of this configuration directly affects the viability of the investment.

Due to the large number of variables involved in the design of an offshore wind farm, and the nonlinear influence they exert on the profitability of the park, it remains impossible to use a method of systematic search for the optimal solution. Therefore, and similar to the case of onshore wind power plants, the optimization methods does not accomplish an exhaustive search, and they fail to guarantee the optimal solution being attained. However, the problem definition, and therefore how to deal with the solution, presents significant differences with the onshore case. These distinctive differences have not been fully taken into account so far by commercial optimization applications like *Resoft WindFarm* or *DNV GL's WindFarmer*, and moreover they offer a robust but closed software where researchers cannot include their own variations.

Indeed, in the case of onshore locations, where it is possible to install smaller parks, the preferred strategy uses genetic algorithms [1] with a previous discretization of the available area. These algorithms employ a codification (genotypes) of the possible solutions, as in [2,3]. Since the grid dimension is defined and fixed prior to execution of the algorithm, the constraint of the position of the turbines to the centre of their cells impedes the assessment of the possibility of variation of these distances. When the number of turbines of the wind power plant increases, the number of possible solutions increases drastically, and hence, unless guarantees to successfully finding the optimum are relaxed, the time needed to reach the optimum increases by the same ratio.

As an alternative, in this paper we have chosen to address the previously unavoidable limitation of the field of solutions by proposing that the turbines follow a regular pattern [4]. This is coherent with the fact that, to date, most offshore wind farms are set in regular pattern layouts in order to reduce their visual impact and to ensure navigation in the area. Initially, only the rhomboid-shaped pattern is assessed, defined by a set of eight parameters whose values establish the solution to the problem.

This change in the structure of the solution has allowed using a non-discrete evolutionary algorithm. It uses continuous values for the solutions of the problem, instead of coded/discrete solutions used in genetic and discrete evolutionary algorithms, as in [5,6]. In a more advanced work, Réthoré et al [7] presents an optimization framework that can use sequential linear programming or simple genetic algorithms, but still limits the possible layout candidates to a list of fixed points inside the domain boundaries. In this work a gradient-based search is used to speed up the algorithm.

The definition of a search engine based on continuous evolutionary algorithms is a simple task, although it has not been used in wind farm layout optimization problems [1]. As a consequence, there is a set of operations necessary for the evaluation of the objective function that have not been analysed or collected in studies of wind farms: including areas with different bearing capacity, depth contours with varying foundation costs depending on the depth, and the point of transition to the coast. There is another process improvement operation, such as a local search operator based on the gradient, whose particular problematic is also presented.

The remaining sections of the paper are structured as follows. Section 2 lists the set of costs and prices used for the simulations and the sources that include them. They have been necessary for an initial approach to the required investment and its expected profitability. Section 3 explains the proposed algorithm and the necessary functions for the evaluation of the components that affect the investment profitability, which are specific for the offshore installation. It also summarizes the operators used in the proposed evolutionary algorithm, and the gradient based local search method. In Section 4, the simulation results are presented. Section 5 summarizes the presented work and outlines a set of conclusions.

At the end of the paper, an appendix is included dealing with the interpretation of level curves.

2. Costs in an Offshore Wind Power Plant

The following data have been used for the simulations, obtained from [8]. In order to establish a normalized value (€ at 2016) for each of the items, the costs were converted from the original currency to € at the rate registered in the commissioning year, and then increased according to the accumulated inflation. Such data are realistic but they cannot be considered reliable for a complete and thorough study of the cost of investment and decommissioning. However, since those costs where there is a greater unavailability are fixed costs, such as in grid connection, management, finance and administration, they are not involved in obtaining the best site.

Design and project management: 95 k€/MW.

Vessel mobilization and demobilization: 880 k€ for the lease of two vessels, located at a distance of 500 km from the holding port.

Turbine cost: 900 k€/MW. It comprises the acquisition cost (AC = 85%), the shipping and assembling (SA = 5%), and the electrical installation (EI = 10%) [9].

Foundation cost: 450 k€/MW referred to 15 m depth, with an increase of 2% per metre depth. Regarding the type of soil, the following types have been considered: zone 1, with no increase in the foundation cost; zone 2, with an increase of 30%; zone 3, with an increase of 60%; and zone 0 or forbidden areas, with a very high cost increase.

Acquisition cost of inner array cables, export cables and onshore HV cables: see subsection 2.1.

Installation cost of inner array cables and export cables: 120 €/m and 170 €/m respectively.

Excavation and installation of onshore HV cable: 400 €/m.

Offshore substation: 95 k€/MW.

Power factor compensating devices and shoreline transitions: 128 €/MVA.

Onshore substation: 60 k€/MW.

Connection to the grid: 200 k€/MW.

Operation and maintenance costs: 15 €/MWh with an annual increase of 5%.

SCADA system: 50 k€/turbine.

Decommissioning costs and residual price: 120 k€/MW.

Price of MWh and expected increase: 90 €/MWh with an annual increment of $\Delta p_{MWh} = 4\%$.

2.1. Cables

2.1.1.   Inner array cables

Table 1 has been obtained from information provided by [10] with respect to two manufacturers, designated as A and B. Original prices were given in $\$_{2007}/m$, and are normalized to $€_{2016}/m$. These data have been used for the selection of the most suitable cable in terms of the purchase cost and the energy losses over the lifespan of the installation.

A supplementary cable extension of 40 m has been added to each turbine for connections.

Table 1. Acquisition cost of inner array cables.

Cross Area (mm^2)	Fixed losses (W/m)	Variable losses $(mW/m/A^{1/2})$	I_{max} (A)	Norm. Price ($€^{2016}/m$)
A95	0	0.714	380	128
A150	6	0.435	430	192
A400	24	0.192	680	321
A630	34	0.123	780	481
A800	50	0.086	900	506
B95	0	0.833	260	384
B150	6	0.500	360	417
B400	8	0.172	640	514
B630	10	0.111	790	535
B800	12	0.086	900	616

2.1.2.   Export cables and onshore cables

Table 2, lists the information for 3-core 220 kV subsea cables, and for 1-core copper cables. With regard to the subsea cable, obtained from [11], a J-tube de-rating factor of 0.88 has been applied. The data for the onshore cable (XLPE insulated, corrugated aluminium armoured cable), has been obtained from a manufacturer.

Table 2. Acquisition cost of export and HV onshore cable.

| Voltage | Section | Export cable | | Onshore cable | |
| | | Capacity | Norm. Cost | Capacity | Norm. Cost |
(kV)	(mm^2)	(MVA)	($€_{2016}$/m)	(MVA)	($€_{2016}$/m)
220	500	250	844	273	232.8
220	630	273	946	297	266.0
220	800	295	1061	314	299.3
220	1000	314	1214	348	367.3

3. Objective and Methodology

The ultimate goal of this article is to provide the knowledge necessary to program a complete algorithm to obtain the optimal layout of an offshore wind power plant. As a starting point, we impose a rhomboidal shape for the plant, composed of a series of clusters or arrays, each grouping the same number of turbines (see Figure 1). This configuration allows a drastic reduction in the time spent on calculating the power loss due to the wakes, which constitutes, by far, the most time-demanding task on each iteration.

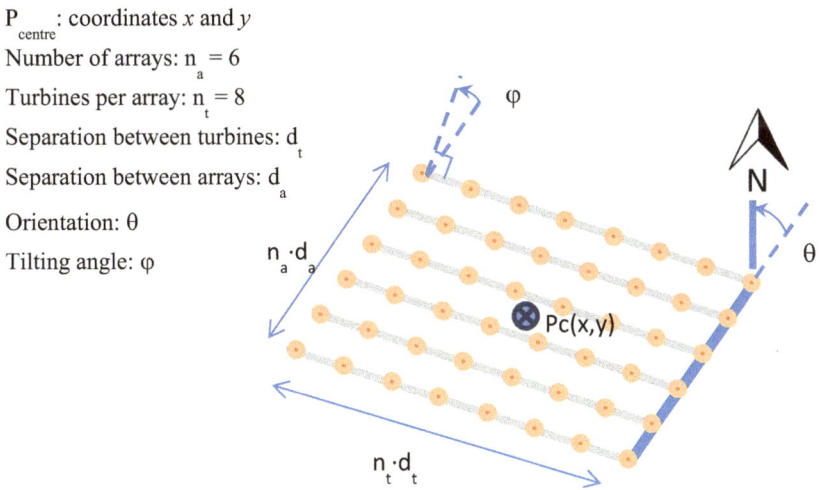

Figure 1. Possible solution for the layout optimization problem following a regular pattern.

Figure 1 shows a possible solution to the search algorithm. The configuration to be optimized is determined by eight parameters whose values define a possible solution to the problem: number of arrays, n_a ; number of turbines per array, n_t ; distance between arrays (expressed in turbine diameters), d_{ba}; distance between turbines in the same array (in diameters), d_{bt}; orientation with

respect to the North θ; the two coordinates of the centre position, $P_c(x, y)$; and tilting angle,

φ (which is null for the particular case of a rectangle). All the values have a continuous variation
range, except for the number of arrays, and the number of turbines per array.

For the sake of operativeness, these last two parameters have been replaced with another two

ones: total number of turbines n and quadrature ratio, defined as $qt = tan^{-1}(n_a/n_t)$. This allows

certain operators (local search and mutation) to increase the overall number of turbines, n, in one
unity in order to explore solutions close to the original one.

To accomplish this objective, a series of partial tasks must be performed. They are presented in
the following subsections, although we can anticipate that they can be represented in the flow chart
of Figure 2.

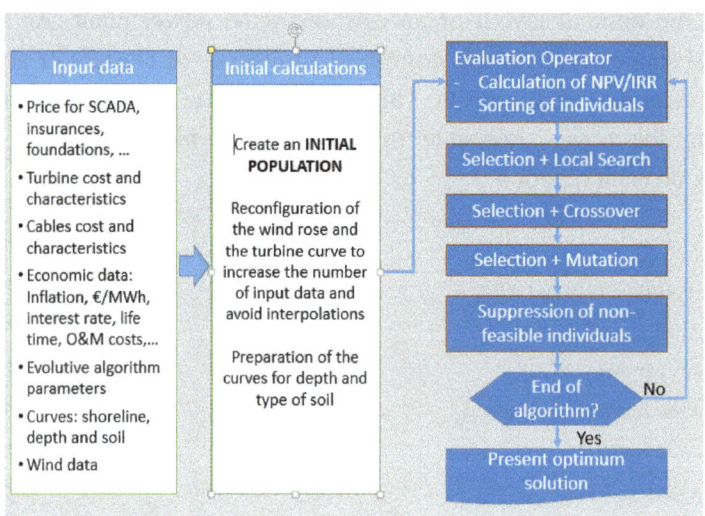

Figure 2. Flow chart for the optimum search algorithm.

3.1. Objective function

In a first step, the Net Present Value (NPV) was used as Objective Function (OF), as a figure of
the investment profitability. NPV is calculated from all economic data relating to the initial cost of
investment, operation and maintenance costs, and the income derived from the sale of electricity [3].
It also must include the decommissioning cost, and the residual value of the installation.

However, two main drawbacks have been observed when analysing the simulation results:

• The NPV is a dimensional value and, by itself, does not express a degree of profitability.
Another figure, typically the initial investment, must accompany the NPV.

• The solutions tend to include excessive investment since an increase in expenditure is
accepted if the NPV is slightly increased after the total lifespan.

For these reasons, the internal rate of return (IRR) is finally selected as the objective function. One of the traditional features of this figure lies in its role of investment quality indicator, in contrast to that of the NPV which strives to increase the shareholders' wealth. However, in the offshore wind farm case, where the economies of scale are strongly present, a high IRR can only be obtained with high investment, which leads to similar solutions to those pursuing an optimum NPV.

The main disadvantage to optimizing the IRR is the fact that the inner-array cables are selected by means of an NPV analysis, and therefore two indicators are used in the overall optimization process.

3.2. Non-discrete evolutionary algorithm

The search engine is based on a non-discrete evolutionary algorithm whose objective is the maximization of an OF. The algorithm does not use a binary encoding for each parameter but tracks its variability interval in a continuous way. As a consequence, crosses between individuals will be made as weighted averages, and mutations as random assignment. It has also allowed including a local search operator based on the OF gradient.

The evolutionary algorithm starts from an initial population of possible solutions to the problem of wind farm layout optimization. To this end, it assigns random values to each of the parameters involved in the configuration of the wind farm, which are presented in Figure 1. The algorithm then enters an iterative process with the following operators (see Figure 2):

Evaluation operator

This operator calculates the OF for each individual in the generation. The OF is a measure of profitability, in this case the NPV or the IRR. Following, it orders the individuals based on their OF, and selects the best n, where n is the population size (input data).

Selection operator

An individual is pseudo-randomly chosen: in half of the situations, the roulette wheel selection is used, in which those individuals with better OF constitute the fittest individuals with an increased chance of being selected; in the other half, a random individual is chosen. This operator is used for the creation of new individuals via the following three operators.

Local maximum operator

This is shown in subsection 3.4. crossover operator.

Two individuals are selected that behave as parents, and two children are created from them. Their characteristics can be selected, in a random way: either by randomly interchanging the parents' characteristics; or from the average weights of the parents' characteristics.

Mutation operator

A single parent is selected, and a new feasible individual is created in which, for every characteristic, the operator randomly chooses between either copying the parent's characteristic, or randomly deducing a value within the permitted range.

Suppression of non-feasible individuals

There is a high probability that one of the turbines of the newly created individuals lies in a forbidden zone, onshore or outside the search space. Inclusion of the suppression operator as part of the crossing and mutation operators therefore leads to a better result. These operators also suppress an individual if the number of turbines exceeds the permitted one.

3.3. Interpretation and use of input data

3.3.1.   Interpretation of the curve defining the shore contour

The set of points that define the outline of the coast must be given as a part of the input data. Solutions with turbines onshore should be discarded. To this end, appendix 6 indicates an efficient and simple-to-program method to check if a point lies within the closed curve.

The same or a similar procedure is also used for the interpretation of the forbidden zones, areas with different bearing capacity, and depth curves.

3.3.2.   Interpretation of the forbidden zones

The set of forbidden zones is comprised of the prohibited zones due to ecological reasons or maritime transport. Each forbidden zone is a closed curve characterized by a set of points. In the case when the algorithm locates any turbine in a forbidden zone, the individual must be suppressed.

3.3.3.   Interpretation of the curves of soil bearing capacity

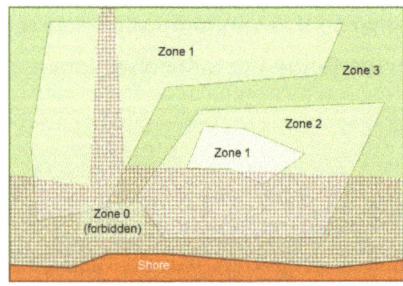

Figure 3. Forbidden zones and curves delimiting the type of sea-bed as a function of its load bearing capacity. Zone 0 = forbidden; zone 1 = reference foundation cost; zone 2, overcost = 30%; zone 3, overcost = 60%.

In order to obtain the bearing capacity at a point $P(x, y)$ of the map, it is necessary to perform a set of operations, some of which can be made before launching the iterative search. Other operations will be specific to the position P, and therefore they must be performed at every turbine location of each individual (see Figure 3):

• As a first step before entering the loop

(a) Read the points of the m curves that define the map of the several types of soil, and their bearing capacities.

(b) Fill in a matrix M_s in which the element $m_s(i, j)$ indicates whether the curve i is within the closed curve j.

• Within the iterative process, for each $P(x, y)$

(a) Obtain the set of curves for which the point $P(x, y)$ lies within.

(b) From among these closed curves, select from matrix M_s the polygon which lies within all the others polygons.

(c) Obtain the bearing capacity of this curve.

3.3.4.   Interpretation of depth curves

The interpretation of depth curves is greatly simplified if these curves are introduced in such a way that, on running their points sequentially, increasing depths are marked on the right (see Figure 4).

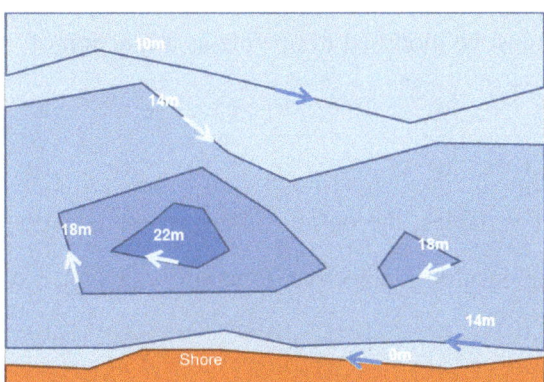

Figure 4. Nautical chart expressing the depth in metres. When travelling along the polygon points, the increasing depths are on the right.

In order to obtain the depth at a point $P(x, y)$, it is necessary to carry out the following:

• As a step previous to entering the loop

1. Read every point of the m polygons that define the nautical chart, and their depth.

2. Obtain whether each curve is travelled CW (leaving inside the decreasing values of depth), or CCW (decreasing values). Appendix 6 deals with this problem.

• Within the iterative process, for each $P(x, y)$

3. Calculate the minimum distance from point P to each curve, and put the curves in order according to their distance.

4. Deduce if the point is at the left or at the right of the closer curve C_{cl0} (see Appendix 6).

5. If it is at the right (analogously left), travel the remaining curves in an increasing order of proximity to P, and select the first curve leaving P at its left (right). This curve is designed as C_{cl1}.

6. Obtain the depth of P as a function of the curve depths C_{cl0} and C_{cl1}, and their distances to P.

3.4. Gradient-based local maximum operator

A local maximum operator is incorporated, which tracks in the vicinity of a provisional solution in order to obtain the set of values that maximizes the objective function.

Thus, within a generation, the algorithm selects the best individual of each generation, and an additional but randomly chosen individual, and they are both then subject to a local optimization process. This operator is based on the local maximum gradient method for continuous functions. In order to prevent any loss of coherency of the gradient method when applied to dimensional variables, the coordinates must be non-dimensionalized by dividing each coordinate by its corresponding allowable range. This way, if $f^{orig}(X_1, X_2, \ldots, X_n)$ is the original objective function to be optimized, this function must be modified to operate as a function of $X = (x_1, x_2, \ldots, x_n) \in \mathbf{R}^n$, with

$$x_j = \frac{X_j}{range(X_j)}$$

With this change of variables, the optimization process consists of detecting the direction of maximum growth of the resulting objective function $f(x_1, x_2, \ldots, x_n)$, which is parallel to its gradient.

$$X = (x_1, x_2, \ldots, x_n) \in \mathbf{R}^n, \quad \Delta_{max\_growth} X \| \nabla f(X) \tag{1}$$

In this case, maximizing the function $f(X)$ starting from an individual with components $X = (x_1, x_2, \ldots, x_n)$ entails:

1. For each component j:

Check if varying the component j by an amount $\Delta^{try} x_j > 0$ leads to a feasible individual. If so, calculate the relative increase $\Delta f(X)/\Delta^{try} x_j$. If it is not feasible, or there is a decrease in $f(X)$, assign $\Delta f(X)/\Delta^{try}x_j = 0$.

Perform the operation above, but this time for $\Delta^{try}x_j < 0$.

Take $\nabla_j f(X)$ as the relative increase with the higher absolute value.

2. Deduce the component *opt* with the higher function increase $\nabla_{opt} f(X) \geq \nabla_j f(X) \; \forall j \leq n$. If it is 0, it is already a local maximum, and the operation ends.

3. Increase every component j by the amount

$$\Delta x_j = \frac{|\Delta^{try}x_{opt}\nabla_{opt} f(X)|}{\nabla_j f(X)} \tag{2}$$

The values of the increments $\Delta^{try}x_j$ are assigned heuristically. These values should be kept relatively high to speed up the algorithm. Another effect (neither entirely positive nor negative) is the higher probability to escaping from one local maximum to another, and therefore potentially optimum solutions can be inadvertently skipped. This issue cannot be seen as a limitation in the design, since a skipped potentially optimum solution implies that the configuration may be excessively sensitive to the input parameters.

Inclusion of this operator produces an effect similar to particle swarm. It introduces a certain guiding of the individuals in the direction of the gradient of the objective function [12]. The gradient vectors act as the particle swarm velocity vectors, although in the proposed algorithm, the vectors are sporadically applied: when crossing operator creates new individuals with similar characteristics to that of individuals locally optimized (see Figure 5). In this way, part of the local optimization performed over the best individual ($parent_1$ in the figure) is transmitted to its sons.

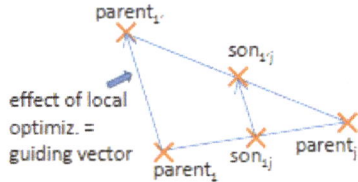

Figure 5. Local search optimization producing an effect similar to the guiding vectors of particle swarms. Individuals considering the local optimization are denoted as primed.

3.5. *Optimized methods for the energy deficit calculation due to wakes*

As previously mentioned, grouping the turbines in a wind park gives rise to a reduction in the produced energy due to the perturbations of upwind turbines in the direction of the flow stream.

In general terms, the net energy generated by the power plant is calculated after evaluating the power deficit that each upwind turbine occasions in each downwind turbine. Furthermore, this calculation must be repeated for all wind speeds considered, and for all wind directions considered. It is possible to implement an improved calculation method to reduce the necessary computation time by a ratio of up to $1/300$ with regard to traditional methods.

To this end, a set of operations must be applied:
• Storing the result of certain operations (mainly the position of the individual turbines).
• Disregarding the wake effect between two turbines placed very far apart.
• Disregarding wind directions greatly deviated from the direction that links two turbines.
• Disregarding the wake effect at very high speeds.
• Suppression of unnecessary calculations where no perturbation is expected.

Since the restriction referring to the rhomboidal shape of the wind part is imposed, a set of operations that take advantage of this regular pattern can be done to reduce even more the computational time [13]. These operations lead to an optimized energy calculation method that reduces the computational time by a factor of $1/20000$ (for a 100-turbine farm).

3.6. Shoreline transition

In order to obtain the linking point between onshore and offshore, the algorithm uses a simplified method that takes into account neither the orography in the sea, nor the orography on the coast. It only calculates the transition point as a function of shoreline contour, the costs of the high voltage cables, and the costs of its installation.

Figure 6 illustrates the concept that supports the calculation of the position of the shoreline transitions. For each section that makes up the shoreline, we must obtain the point belonging to the line containing the segment, where the overall cost is lower. If the point is beyond the interval of coast evaluated, the closest vertex is chosen.

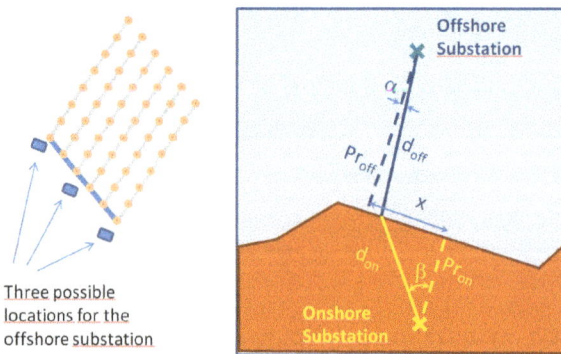

Figure 6. Point of connection between offshore and onshore cables.

Therefore, the objective consists in minimizing

$$d_{off}C_{off} + d_{on}C_{on} = f(d_{on}, d_{off}) \tag{3}$$

where *off* and *on* stand for *offshore* and *onshore*, *d* is the distance to the transition point and *C* represents the costs of purchasing and installing the cable. This expression can be rewritten as

$$C_{off}\sqrt{Pr_{off}+(X-tPr_{on})^2}+C_{on}Pr_{on}\sqrt{1+t^2} \qquad (4)$$

where $t = \tan(\beta)$, and *Pr* stands for projection (see Figure 6). Minimizing this expression for *t* is not trivial, and it requires an iterative process to obtain the root of its derivative.

This calculation must be performed for each of the 3 possible locations of the offshore substation with regard to the wind farm (see Figure 6), and the location with the lowest cost (including the substation platform foundation cost) is then selected for the corresponding individual under evaluation.

3.7. Selection of the most suitable section and model for the inner-array cables

The current flowing along an array is higher as long as the cable collects the power from more turbines. In this way, several section cables can be used along the same array, thus reducing the expenditure of the electrical infrastructure.

Starting from the information available regarding cable price, capacity and expected losses, an offline analysis has been performed for each stretch along an array. Therefore, by taking the cable cost and its power losses (normalized into present euros) into account, the most suitable cable can be calculated that gives rise to highest NPV over the lifespan. It is worth mentioning that the IRR cannot be used as indicator because investment and flow cash have the same sign.

4.   Simulation Results

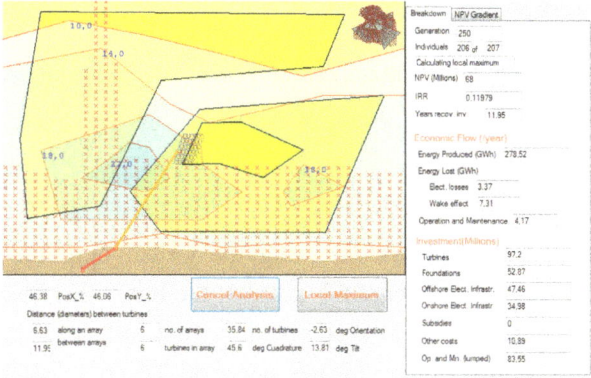

Figure 7. Optimization result. Numerical values for the solution parameters appear in the bottom, and resulting economic data are shown at the right. The wind rose is depicted at the upper-right corner.

An application supported by the introduced methodology, has been designed and programmed to search for the optimum layout and relative location of an offshore wind farm that follows a rhomboidal pattern (see Figure 7).

4.1. Definition of the scenario

The information relative to the scenario is defined by means of an input file, made up of different sheets related to estimated/measured wind, economic figures, turbine characteristics and costs, foundation types and costs, electrical components, definition of curves for depths, soil and coasts, as well as algorithm parameters.

As can be seen in Figure 7 the following scenario has been selected:
• Studied zone of 80 km × 40 km.
• A wind rose given by:

Sector	E	SE	S	SW	W	NW	N	NE
Probability in %	15	20	15	15	5	10	10	10

• Weibull parameters: C param = 8.1 m/s and K param = 2.3.
• The coast is located to the south. The onshore substation is located to the south-west.
• A forbidden strip parallel to the coast, plus an additional perpendicular path for marine sailing is included.
• Three types of soil are considered, the zone 1 being that with highest load-bearing capacity (see Figure 3).
• The sea depth in the administrative concession varies between 10 m and 22 m. The curves are represented for 10 m, 14 m, 18 m (two curves) and 22 m (see Figure 4 and Figure 7).
• There is a limit of 38 turbines, each of 3 MW. Its power curve is given by:

w_s (m/s)	4	5	6	7	8	9	10	11	12	13	14	15	16	17	...	25
P (MW)	0	0.09	0.22	0.43	0.71	1.11	1.55	2.02	2.39	2.77	2.91	2.95	2.98	3.00	...	3.00

where w_s is the wind speed and P is the generated power at this speed.

4.2. Battery of simulation case analysed

Evolutive algorithms are efficient in searching the global optimum for complex problems. However, they fail to guarantee the convergence towards the optimum. For this reason, a set of 300 simulations have been performed over the same scenario. In all of them, the obtained solutions fell into the same configuration: 6 arrays of 6 turbines per array (6×6). Table 3 summarizes the main results. In half of the simulations (identified by $p1$), the maximum number of generations was established as 100, and the population size as 80. In the other half ($p2$) the maximum number of generations was 250, and the population size was 140. The first two columns in this table refer to a set of 100 simulations in which the function objective was the NPV. In the following two columns, simulations refer to an analogous set of simulations in which the function objective was the IRR. The last two rows also refer to simulations with IRR as the objective function, but here the maximum local operator is suppressed.

Table 3. Result of the simulations when using different objective functions. Mean value, standard deviation and best economic indicator are given for the resulting NPV and IRR. Mean value and deviation are given for the computation time and investment.

p1: max gener = 100 popul. size = 80 p2: max gener = 250 popul. size = 140		Objective function = NPV		Objective function = IRR		Objective function = IRR without local search	
		p1	p2	p1	p2	p1	p2
NPV (M€)	Mean	67.10	67.14	67.02	67.16	66.50	66.87
	Deviat.	0.29	0.32	0.30	0.27	0.83	0.33
	Best	67.59	67.61	67.55	67.59	67.24	67.53
IRR (%)	Mean	11.936	11.935	11.937	11.943	11.916	11.931
	Deviat.	0.013	0.015	0.012	0.009	0.031	0.014
	Best	11.958	11.960	11.957	11.959	11.947	11.956
Time (s)	Mean	105.1	220.6	97.9	175.7	104.1	209.3
	Deviat	11.48	71.83	15.45	40.56	14.39	72.67
Investment (M€)	Mean	257.1	257.4	256.7	256.6	256.8	256.8
	Deviat	0.51	0.48	0.53	0.42	0.61	0.43

With regard to the rows, this table organizes the obtained information in four blocks. The first block comprises the average NPV, the standard deviation, and the best NPV for every set of 50 executions. The second block is similar, but referred to IRR. The third and fourth blocks show the mean value and standard deviation for, respectively, the computation time and the required investment. As a reference for the computation time, the algorithm was programmed in managed C++, and executed in a notebook Pentium $2.13GHz$. As a comparison, the running time of 50 cases for an OWF with 20 turbines, presented in [14] was 120 h, while with the presented algorithm, the required time for an OWF with 36 turbines is around 100 s (100 generations).

Table 4 gathers the results obtained from the 300 simulations, showing the mean value and standard deviation of the parameters that define the solutions.

Table 4. Mean value and standard deviation for NPV, IRR, and the parameters that define the 300 solutions to the studied layout problem. Centre in km, separation between arrays in number of diameters, and orientation and tilt angle in deg.

	NPV (M)	IRR (%)	ctr_x	ctr_y	nd_{aa}	nd_{ba}	θ	Tilt
Mean	67.0	11.933	40.9	27.5	6.7	11	- 22	5.4
std	0.49	0.019	0.78	0.56	0.74	1.52	74.2	13.0

The following conclusions can be derived from its analysis:
• Regardless of whether the considered OF is NPV or IRR, similar figures are obtained.
• Slightly lower values of investment are required if the OF is the IRR.
• Logically, simulations designed as *p2* (higher number of generations and individuals per generation) take considerably more time to reach the optimum. However, the results are not

significantly better. Therefore, we recommended performing a higher number of simulations with a lower number of generations and individuals per generation, and then choosing the best solution.

• The local maximum operator increases the mean of the OF for the optimum. It does not entail a significant improvement with regard to the computational time.

In addition to these simulations, further simulations have been performed with higher values for the parameters than those defined for *p2*. However they have not given rise to a solution with a significantly higher OF with respect to those of Table 3.

4.3. Selection of the best optimum

Figure 7 shows the solution with the highest IRR among the 100 obtained for the same scenario (OF = IRR, with local maximum operator). As expected, the centre of the solution, and in fact the centres of all the solutions, fall into the zone with the highest load-bearing capacity (zone 1).

As can be extracted from data on the right-hand-side of Figure 7, the cost breakdown obtained is: turbines (39.9%), foundations (21.7%), electrical infrastructure (33.9%), engineering and others (4.5%), which is similar to that obtained from the reviewed papers summarised in Table 5. However, a significant difference exists in the electrical infrastructure that leads to believe that there may be an overestimation of the cable purchase and laying costs. One of the causes of this possible overestimation lies in the fact that the cable for each stretch has been selected such that its NPV is less negative. This leads to the selection of more expensive cables if it guarantees fewer power losses over the total lifespan. Another cause lies in the increase of the copper price.

Table 5. Breakdown of costs obtained from various sources.

Source	[15]	[16]	[17]	[9]	[18]
Turbines	37%	47%	50%	46%	49%
Foundation	25%	29%	25%	23%	21%
Elect. Infr.	23%	16%	15%	21%	21%
Engineering	12%	7%	5%	10%	9%
Others (SCADA, decommissioning)	3%	1%	5%		

5.   Discussion and Conclusions

The aim of this research is to program a non-discrete evolutionary algorithm to optimize the layout of offshore wind farms. In order to delimit the solution space, the layout has been constrained to a rhomboid-shaped regular pattern. So as to accomplish the programming of the optimum search algorithm, two basic keystones must be solved: obtaining realistic data about the costs and characteristics of the most significant items in the wind farms; and providing systematic and simple methods to manage the various depths and types of seabed.

A major additional functionality has also been explained: a gradient-based local maximum operator. Including this operator improves the IRR values for the optimum, although it does not improve the algorithm convergence (in term of computation time) in the studied cases.

Starting from the knowledge of these functionalities, a non-discrete evolutionary algorithm can be easily programmed and tested. The resulting configuration is the logical one, and is located in the zone with the best conditions of depth and type of soil. The final cost per MW and its breakdown are similar to those given in the literature.

Two possible objective functions have been tested: NPV and IRR. The results are very similar, although the option OF = IRR leads to lower investments.

Taking advantage of the rhomboidal shape, the algorithm uses the method presented in [12] to reduce the required time to compute the energy lossed due to the wake effect. As a consequence, the complete algorithm is able to reach an optimum in approximately 100 s (36 turbines, 100 generations, 80 individuals per generation), drastically faster than existing algorithms.

A. Position of a point with regard to a closed curve

It is necessary to program a function to deduce whether a point is inside or outside a closed curve. This function is employed to ascertain: whether a point is on the coast; whether it is in a forbidden zone; to which load-bearing capacity zone it belongs; and in a modified way, to calculate the depth of the sea bed. As an illustration, an example is laid out at the end of the section.

It is convenient to first define the operation $\overset{\bullet}{+}$ as well as its opposite $\overset{\bullet}{-}$ as

$$a,b \in [-\pi, \pi],$$

$$a \overset{\bullet}{+} b \Leftrightarrow \begin{cases} a+b & if \quad a+b \in [-\pi, \pi] \\ a+b+2\pi & if \quad a+b < -\pi \\ a+b-2\pi & if \quad a+b > \pi \end{cases} \tag{5}$$

$$a \overset{\bullet}{-} b \Leftrightarrow \begin{cases} a-b & if \quad a-b \in [-\pi, \pi] \\ a-b+2\pi & if \quad a-b < -\pi \\ a-b-2\pi & if \quad a-b > \pi \end{cases} \tag{6}$$

The operation $\overset{\bullet}{+}$ together with the interval $[-\pi, \pi]$ constitute an abelian group satisfying commutativity, closure, associativity, existence of neutral element (0), and an opposite element (the opposite of c is $-c$).

A.1 Point inside a closed curve

With these definitions, and assuming that a closed curve L is composed of a sequence of m points (x_i, y_i) (in which $x_m = x_0$ and $y_m = y_0$), then in order to ascertain whether a point $P(x^p, y^p)$ is inside or outside this curve, it is sufficient to evaluate the summatory

$$\Gamma = \sum_{k=1}^{m} \left\{ \arctan(y_{k+1} - y^P, x_{k+1} - x^P) \dot{-} \arctan(y_k - y^P, x_k - x^P) \right\} \qquad (7)$$

By means of Stokes' theorem, it can be demonstrated that

$$P(x, y) \; is \; inside \; L \Leftrightarrow \Gamma = 2\pi \quad or - 2\pi \qquad (8)$$

$$P(x, y) \; is \; outside \; L \Leftrightarrow \Gamma = 0 \qquad (9)$$

With this formulation, then $\Gamma = 2\pi \quad or - 2\pi$ for points belonging to the contour. In order to detect this situation, it is preferable to check each value of the summatory in (7). The point will belong to the contour if and only if any of the addends yields $-\pi$ or π.

A.2 Curve travelled CW or CCW

The point sequence in a curve indicates if the curve is travelled clockwise (CW), or counter-clockwise (CCW).

As previously seen, if a point $P(x, y)$ is inside the curve, then $\Gamma_{inner} = 2\pi \quad or - 2\pi$. If the curve is travelled CCW, then $\Gamma_{inner} = 2\pi$, and the inner points remain on the left of the curve being travelled. Outer points remain on the right.

If the curve is travelled CW, then $\Gamma_{inner} = -2\pi$, and the inner points are on the right of the curve being travelled. Outer points remain on the left.

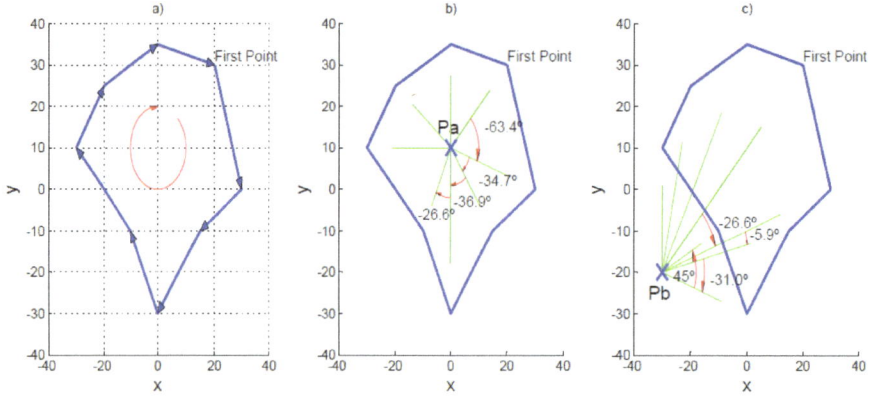

Figure 8. a) Closed curve travelled clock wise; b) differences of angles for an inner point; c) differences of angles for an outer point.

However, if no inner points are known, the way to detect whether a curve is travelled CW or CCW is based on the cross product. Thus, a curve is travelled CCW if the summatory of cross products of all of the consecutive segments is negative, and CW if the result is positive.

A.3 Example

We can assume a closed curve defined by

$$\mathbf{L} \equiv \{(20,30), (30,0), (15,-10), (0,-30), (-10,-10), (-30,10), (-20,25), (0,35), (20,30)\}; \qquad (10)$$

where, for the sake of space, the z component has been suppressed. By drawing it (see Figure 8a), we could anticipate that it is travelled CW, and we can check it by calculating the summatory of the cross products.

$$\sum_{i=2}^{8}(\mathbf{L}_{i+1}-\mathbf{L}_{i})\times(\mathbf{L}_{i}-\mathbf{L}_{i-1})+(\mathbf{L}_{2}-\mathbf{L}_{1})\times(\mathbf{L}_{1}-\mathbf{L}_{8})$$

$$= 550-150+500-200+500++200+300+550 = 2250 > 0 \Rightarrow CW$$

If, as an example, we want to test whether point $Pa(0,10)$ is an inner point or outer point, we can calculate every addend in (7). Before checking whether they belong to $[-\pi,\pi]$, the list of addends is the following (translated to deg):

$$-63.43 -34.70 -36.87 -26.57 \ 296.57 -36.87 -53.13 -45.00 \qquad (11)$$

where the first four values are represented in Figure 8b. After forcing each addend to belong to $[-\pi,\pi]$, this becomes

$$-63.43 -34.70 -36.87 -26.57 -63.43 -36.87 -53.13 -45.00 \qquad (12)$$

Summation of these figures yields -2π. The negative value represents (again) that the curve is travelled CW. The non-null value indicates that point Pa is inside $\mathbf{L}$, and is on the right.

If we want to test the point $Pb(-30,-20)$, then the list of addends is

$$-26.57 -5.91 -30.96 \ 45.00 \ 63.43 -12.53 -16.08 -16.39 \qquad (13)$$

where all figures belong to $[-\pi,\pi]$. As a sample, the first four values have been represented in Figure 8c. After summing them, this yields 0, and therefore, point Pb is outside $\mathbf{L}$, and is on the left.

Conflict of Interest

All authors declare no conflicts of interest in this paper.

References

1. Tesauro A, Réthoré P E, and Larsen G C. (2012) State of the art in wind farm optimization. *Proc EWEA*, 1–11.

2. Mosetti G, Poloni C, and Diviacco B. (1994) Optimization of wind turbine positioning in large wind farms by means of a genetic algorithm. *J Wind Eng Ind Aerod* 51: 105–116.

3. González J, Gonzalez A G, Castro J, et al. (2010) Optimization of wind farm turbines layout using an evolutive algorithm. *Renew Energ* 35: 1671–1681.

4. Nielsen P. (2003) Offshore wind energy projects feasibility study guidelines seawind. *Tech Rep*, Altener Project.

5. Hajela P. (1990) Genetic search-An approach to the nonconvex optimization problem. *Aiaa J* 28: 1205–1210.

6. Serrano J, Burgos M, and Riquelme J M. (2011) An improved evolutive algorithm for large offshore wind farm optimum turbines layout. *Power Tech 2011 IEEE Trondheim*, 1–6.

7. Réthoré P E, Fuglsang P, Larsen G C, et al. (2014) Topfarm: multi-fidelity optimization of wind farms. *Isope* 17: 1797–1816.

8. Gonzalez A G. (2017) Review of offshore wind farm cost components. *Energ Sust Dev*, 10–19.

9. Douglas W. (2010) Offshore wind asessment for Norway. *Tech Rep,* Norway.

10. Green J, Bowen A, Fingersh L J, et al. (2007) Electrical collection and transmission systems for offshore wind power. National Renewable Energy Laboratory (U.S.), 1–7. Available from http://www.nrel.gov/docs/fy07osti/41135.pdf.

11. Offshore development information statement (ODIS). (2009) Appendices. *Tech Rep*, National Grid.

12. Hou P, Hu W, Soltani M, et al. (2015) Optimized placement of wind turbines in large-scale offshore wind farm using particle swarm optimization algorithm. *IEEE T Sust Energ* 6: 1272–1282.

13. Gonzalez A G, Burgos M, Riquelme J, et al. (2015) Reducing computational effort in the calculation of annual energy produced in wind farms. *Renew Sust Energ Rev* 43: 656–665.

14. Pérez B, Mínguez R, and Guanche R. (2013) Offshore wind farm layout optimization using mathematical programming techniques. *Renew Energ* 53: 389–399.

15. Offshore Design Engineering (ODE) Limited. (2007) Study of the costs of offshore wind generation. Tech Rep URN number 07/779, Renewables Advisory Board (RAB) & DTI. Available from http://webarchive.nationalarchives.gov.uk/+/http:/www.berr.gov.uk/files/file38125.pdf.

16. Morgan C A, Snodin H M, and Scott N C. (2003) Offshore wind economies of scale, engineering resource and load factors. *Tech Rep*, Garrad Hassan.

17. Slengesol I, De Miranda W P, Birch N et al. (2010) Offshore wind experiences: a bottom-up review of 16 projects. *Tec Rep*, OceanWind.

18. Wind Energy-The facts. (2009) Economics: the cost of energy generated by offshore wind power. *EWEA*. Available from https://www.wind-energy-the-facts.org/images/chapter3.pdf.

Effect of the storage condition of microalgae on hydrochar lipids and direct esterification-transesterification of hydrochar lipids for biodiesel production

Vo Thanh Phuoc * and Kunio Yoshikawa

Yoshikawa Laboratory, Department of Environmental Science and Technology, Tokyo Institute of Technology, Yokohama, Kanagawa, 226-8502, Japan

* **Correspondence:** Email: vo.p.aa@m.titech.ac.jp.

Abstract: The hydrochar product from the hydrothermal carbonization (HTC) of microalgae contains most of fatty acids (FAs) in the original microalgae. In the hydrochar, FAs exist in both types of bound fatty acids (BFAs) and free fatty acids (FFAs). Besides, when the microalgae paste is stored at the room temperature (25 °C) for one day, there is an increase of total fatty acids (TFAs) and free fatty acids (FFAs) in microalgae. The hydrochar from this microalgae paste was proved to have a higher amount of TFAs and a higher percentage of FFAs/TFAs compared to the ordinary hydrochar (without the additional storage step) in this research. Both of these factors favor for the subsequent acid catalyzed esterification-transesterification reaction of hydrochar lipids. In summary, a process based on a combination of the storage of fresh microalgae, the HTC of microalgae paste, and the direct esterification-transesterification of the hydrochar has been developed for biodiesel production. With the additional storage step of fresh microalgae, the total biodiesel yield has been improved of 19.3% in the optimum condition.

Keywords: microalgae; hydrothermal carbonization; hydrochar; fatty acid; esterification; transesterification

1. Introduction

Recently, biodiesel from microalgae is especially concerned because microalgae can grow very fast, have high lipid content and can be cultivated at non-arable land or combined with sewage treatment [1]. However, due to very high moisture content of microalgae paste (approximately 80–90% w/w), drying of microalgae consumes so much energy. Furthermore, the treatment of a large amount of byproduct (lipid-extracted microalgae) is also another obstacle. One solution to these problems is the hydrothermal carbonization (HTC) process.

HTC of microalgae is a process in which microalgae react with water at high temperature (around 200 °C, 2 MPa) to form one solid product called hydrochar. Treatment of microalgae by HTC is very beneficial in terms of energy. For 10 kg microalgae paste with 90% moisture content, the energy required for complete drying is 23.31 MJ, while the energy required for performing HTC (without evaporation of water) and drying of the hydrochar is 7.83 MJ, only around 1/3 of the original value [2]. The heat of combustion of the lipid-extracted hydrochar was proved enough to provide the whole energy requirement of the HTC process and make the energy balance positive [2,3]. In comparison with a similar process of the hydrothermal liquefaction, HTC takes place in the milder condition, so it means that the energy requirement and the equipment cost will be considerably reduced. For example, in the hydrothermal liquefaction, a very high temperature (375 °C) should be required to get the maximum amount of bio-crude oil. Moreover, the bio-crude oil from the hydrothermal liquefaction consists of many various substances such as hydrocarbons, cyclic dipeptides, pyrroles, pyrrolidinones, free fatty acids, peptides, algaenan (polymethylens), algaenan derivatives, asphaltene, etc., has high viscosity and percentage of nitrogen, so these will add more cost to upgrade this oil before utilization [4]. On the other hand, after HTC of microalgae, two main products are obtained. While the aqueous phase, which contains the majority of N in the original microalgae, can be reused for cultivating microalgae [3,5], the hydrochar contains most of fatty acids, and it is a promising feedstock for biodiesel production [2,3].

Fatty acids in the hydrochar are bound fatty acids (BFAs) and free fatty acids (FFAs) as a result of the hydrolysis of BFAs during HTC [6]. One of the most effective methods to produce FAMEs from the hydrochar is a direct esterification-transesterification. Fatty acids (FAs) inside the hydrochar (both BFAs and FFAs) will directly react with methanol in the presence of a catalyst to form fatty acid methyl esters (FAMEs) without oil extraction. This approach eliminates a big amount of hazardous solvents required for oil extraction from microalgae such as chloroform and methanol [7], hexane and isopropanol [8]. Moreover, the biodiesel yield will be also improved compared to the traditional methods including the oil extraction and the subsequent transesterification [9,10,11]. Acid catalysts, the commonly used one is H_2SO_4, are usually used to catalyze both esterification of FFAs and transesterification of BFAs in one step.

In addition, the contents of TFAs and FFAs in microalgae paste increase with the storage time. In detail, there is an increase of FFAs along with a decrease of triacylglycerides (TGs, a form of BFAs) due to the hydrolysis of TGs under the effect of the enzyme lipase in microalgae [12,13]. The microalgae paste should undergo the HTC process to obtain the hydrochar used as a feedstock for biodiesel production. If the TFA content of microalgae is higher, the TFA content of the created hydrochar will be higher at the same HTC condition. However, if the FFA content of microalgae is higher, more FFAs will be extracted into the aqueous phase during the HTC due to the high solubility of FFAs in water at a high temperature [14]. Although most of these FFAs will be adsorbed again on hydrochar [3], a small part will be freely dissolved in the aqueous phase or adsorbed on the dissolved organic matters [15]. The higher the FFA content of microalgae is, the higher the possibility of losing these FFAs in the aqueous phase is. In other words, the TFA retention in hydrochar will be lower. There are two opposite effects on the TFA content of the created hydrochar. Therefore, in this work, HTC of stored microalgae has been conducted to see whether the amount of TFAs in the hydrochar increases compared to the hydrochar from microalgae without storage. In addition, the percentage of FFAs/TFAs is also evaluated in both cases. Because the acid catalyzed esterification of FFAs is much faster than the acid catalyzed transesterification of BFAs, the higher percentage of FFAs/TFAs will make the biodiesel formation faster at the same reaction condition. This additional storage step does not require chemicals or much energy, so it should be highly applicable.

2.   Materials and Methods

2.1. Materials

2.1.1.   Microalgae cultivation, harvest and storage

The marine microalgae species, Chlorella (PTCC 6010), was cultivated in the Rudic medium with the total volume of 120 liters [16]. The medium recipe was as follows, $NaNO_3$: 300 mg/l, K_2HPO_4: 80 mg/l, KH_2PO_4: 20 mg/l, $MgSO_4.7H_2O$: 10 mg/l, $CaCl_2$: 47 mg/l, EDTA: 7.5 mg/l, NaCl: 20 mg/l, H_3BO_3: 0.3 mg/l, $MnSO_4.H_2O$: 1.5 mg/l, $ZnSO_4.7H_2O$: 0.1 mg/l, $(NH_4)_6Mo_7O_{24}.4H_2O$: 0.3 mg/l, $CuSO_4.5H_2O$: 0.08 mg/l, $Co(NO_3)_2.6H_2O$: 0.26 mg/l, $FeCl_3.6H_2O$: 17 mg/l. The salinity of the medium was 30 g NaCl/l. The temperature was remained around 25 °C, and pH was adjusted around 8 by gaseous CO_2. The microalgae density was determined by the UV-Vis spectrometer, and the microalgae medium reached to the stationary phase after 2 weeks and was ready for harvest. Microalgae were flocculated by $FeCl_3$ with an amount of 0.3 g/l of the culture medium [17]. The flocculated microalgae were washed 2 times with water and then filtered to get microalgae paste.

The microalgae paste was divided into 3 parts after harvesting. The first one was dried immediately at 105 °C for 4 hours. The second and the third were stored in the incubator at 25 °C respectively for 1 and 2 days, and then dried at 105 °C after finishing the storage. Drying of microalgae enables the storage of microalgae and the accurate control of microalgae mass input in each experiment. In HTC, a certain proportion of dried microalgae and water was used to simulate the real moisture of microalgae paste. This protocol was just applied for the lab scale, and a fresh microalgae paste should be used for HTC in a commercial scale. The dried microalgae were also used in HTC in many previous researches [2,6,18].

2.1.2.   Chemicals

All chemicals used in the microalgae cultivation, the harvesting and the direct esterification-transesterification reaction were purchased from Wako Pure Chemical Industries, Ltd. The fatty acid methyl ester standards including myristoleic acid, methyl ester (C14:1); tetradecanoic acid, methyl ester (C14:0); pentadecanoic acid, methyl ester (C15:0); 9-hexadecenoic acid, methyl ester, (Z) (C16:1); hexadecanoic acid, methyl ester (C16:0); heptadecanoic acid, methyl ester (C17:0) were purchased from Sigma-Aldrich.

2.1.3. Equipment

The UV-Vis spectrometer model was Shimadzu UVmini 1240. The 50 ml autoclave was purchased from Taiatsu Techno Corporation. The Tomy Multipurpose Refrigerated Centrifuge EX-126 was used to centrifuge samples. The Shimadzu GCMS-QP2010 SE was used to quantitatively analyze fatty acid methyl esters.

2.2. Hydrothermal carbonization of microalgae

HTC of microalge was conducted using the facility shown in **Figure 1**. Dried microalgae (3 g) and water (17 g) were supplied into the 50 ml autoclave. The ratio of microalgae and water was calculated to simulate the microalgae paste with 85% moisture content. One magnetic bar was used to agitate the mixture during HTC at 180 rpm. Next, the mini-autoclave was closed, and the argon gas was used to flush out the air inside.

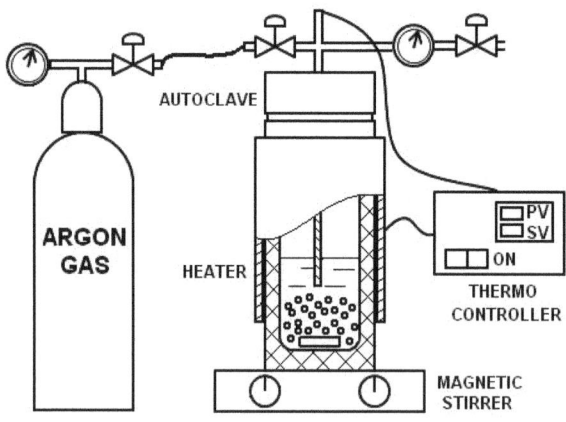

Figure 1. Hydrothermal carbonization of microalgae in the autoclave.

The autoclave was heated to the desired temperature by a heating band and was held for 30 minutes. Preliminary experiments showed that when the HTC temperature was 190 °C or lower, or the holding time was 15 minutes, it was quite difficult to separate the hydrochar from the aqueous phase by filter papers. The reason is that the hydrochar size is not large enough. Hence, HTC was conducted at 200 °C, 210 °C and 220 °C, while the holding time was set at 30 minutes. This range of the HTC temperature and the holding time are appropriate for the HTC condition of microalgae [2,18,19]. Because there are no lignin, cellulose and hemicellulose in the microalgae composition, under a moderate temperature (around 200 °C) and time (< 1 h), microalgae can be converted into hydrochar [2]. After the holding time was reached, the autoclave was cooled down to the room temperature. The temperature profiles of the HTC experiments are shown in **Figure 2**.

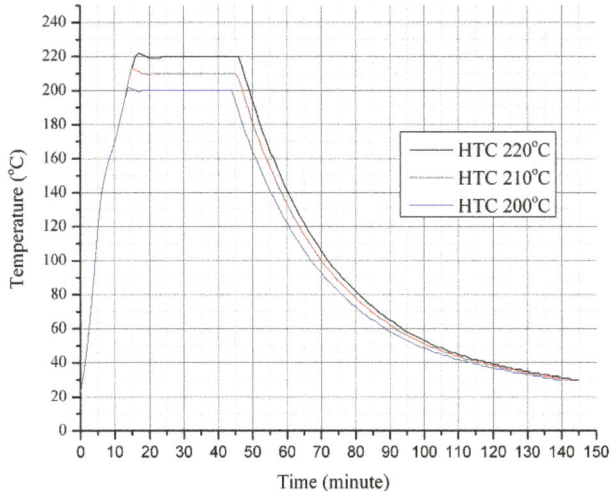

Figure 2. The temperature profiles of HTC experiments.

The product mixture was filtered by the pre-weighed GF-A paper with the pore size of 1.6 μm, and then the hydrochar was washed with 50 ml of distilled water. Next, the hydrochar and the filter paper were dried at 105 °C for 4 h. After that, the char was weighed and stored at 4 °C before conducting the esterification-transesterification experiments.

2.3.   Direct esterification and transesterification of hydrochar lipids

The direct esterification and transesterification of hydrochar lipids were performed in a 10 ml glass tube shown in **Figure 3**. Firstly, the hydrochar (0.1 g or 0.2 g) and a magnetic bar were gently put into the glass tube, and the samples were prevented from sticking on the top of the tube. An amount of 1 ml methanol solution containing H_2SO_4 and the internal standard hexadecanoic acid methyl ester (C17: 0) of 100 μg was then added, and the tube was sealed by a teflon cap. The tube was then placed in a pre-heated silicone oil bath and underwent the esterification and the transesterification at 85 °C for 90 min with the stirring speed of 120 rpm. To reduce the amount of methanol, the maximum amount of hydrochar (0.2 g) was used corresponding to 1 ml of methanol, so two levels of 0.1 g and 0.2 g hydrochar/ml methanol were investigated. Sulfuric acid concentrations were investigated at 1, 2, 3 and 4% v/v of methanol. The temperature of the esterification-transesterification reaction was fixed at 85 °C, similar to the experiments for determining TFAs of microalgae [20].

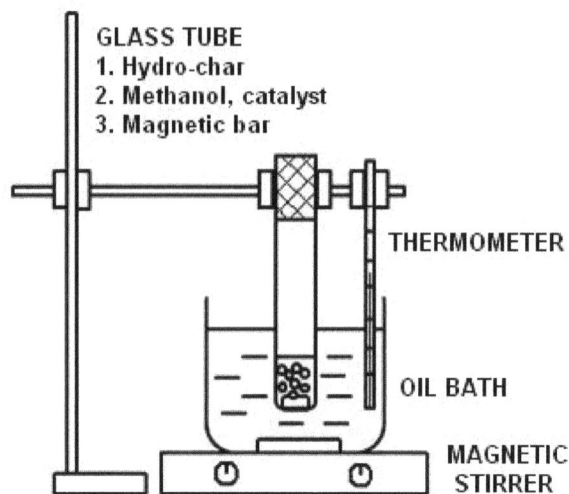

Figure 3. Direct esterification-transesterification of hydrochar lipids in the glass tube heated by the oil bath.

When the reaction time was reached, 1 ml of distilled water was added to stop the reaction. Subsequently, an amount of 5 ml hexane was added into the tube to extract FAMEs. The mixture was shaken strongly and centrifuged at 2000 rpm for 5 minutes to completely separate the aqueous phase and the hexane phase. An amount of 1 ml hexane phase was pipetted to vials which were stored at 4 °C before the GC-MS analysis.

2.4.   Quantitative analysis of total fatty acids (TFAs) and free fatty acid (FFAs) by GC-MS

2.4.1.   Sample methylation for analysis of total fatty acids (TFAs)

The method determining the TFA content of microalgae (or hydrochar) was also based on the direct esterification and transesterification. Only 10 mg of microalgae (or hydrochar), instead of 0.1 g or 0.2 g, was used in this case, and the H_2SO_4 concentration was 2% v/v of methanol. The procedure can be referred above.

2.4.2.   Sample methylation for analysis of free fatty acids (FFAs)

To determine the amount of FFAs of microalgae (or hydrochar), N, N-Dimethylformamide dimethyl acetal was used to selectively convert FFAs into FAMEs [21]. According to the Floch method, a mixture of 0.6 ml chloroform/methanol (2: 1, v/v) was utilized to entirely extract lipids of 10 mg microalgae (or hydrochar) [7]. The solvent mixture of chloroform/methanol was then evaporated completely at 60 °C. Next, 1 ml mixture of pyridine/N, N-Dimethylformamide dimethyl acetal (1: 1 v/v) were added to selectively convert FFAs into FAMEs at 60 °C for 15 minutes. After that, 4 ml hexane containing the internal standard hexadecanoic acid methyl ester (C17: 0) of 100 µg was added. Finally, the mixture was centrifuged at 2000 rpm for 5 minutes, and 1 ml solution was taken into vials which were stored at 4°C prior to the GC-MS analysis.

2.4.3. Quantitative analysis by GC-MS

FAMEs were analyzed by GCMS-QP2010 SE, Shimadzu, with the column Rtx-5MS, 30 m length, 0.25 mm diameter, 0.25 μm film thickness, and the splitless mode. The injector temperature was 300 °C, and the ion source temperature was 200 °C. The temperature program was as follows. The oven temperature was held at 40 °C for 3 minutes, increased to 180 °C at the rate of 20 °C /min, increased to 200 °C at the rate of 5 °C /min, held at 200 °C for 10 minutes, and increased to 300 °C at the rate of 20 °C /min. By using the internal standard method, the component concentrations were quantified by comparing the peak area ratio of each component with the peak area ratio of standard components. The concentration of FAMEs was the total of each component concentration.

2.5. Definition

Microalgae A: fresh algae, microalgae B: one-day-stored algae, microalgae C: two-day-stored algae
Hydrochar A from microalgae A, hydrochar B from microalgae B
Mass yield of hydrochar H_A, H_B
H_A, % = (mass of hydrochar A × 100)/mass of treated microalgae A
H_B, % = (mass of hydrochar B × 100)/mass of treated microalgae B
M_A: TFAs of microalgae A (mg/g algae)
N_A: TFAs of hydrochar A (mg/g hydrochar)
Y_{A1}: TFA retention of hydrochar A, %

$$Y_{A1} = N_A \times H_A/M_A, \%$$

Y_{A2}: yield of esterification and transesterification reaction of hydrochar A, %
Y_A: total biodiesel yield of the whole process from microalgae A to biodiesel

$$Y_A = M_A \times Y_{A1} \times Y_{A2} \text{ (mg/g algae)}$$

M_B: TFAs of microalgae B (mg/g algae)
N_B: TFAs of hydrochar B (mg/g hydrochar)
Y_{B1}: TFA retention of hydrochar B, %

$$Y_{B1} = N_B \times H_B/M_B, \%$$

Y_{B2}: yield of esterification and transesterification reaction of hydrochar B, %
Y_B: total biodiesel yield of the whole process from microalgae B to biodiesel

$$Y_B = M_B \times Y_{B1} \times Y_{B2} \text{ (mg/g algae)}$$

3. Results and Discussion

The major FAs of microalgae A are tetradecanoic acid, methyl ester (C14:0); 9-hexadecenoic acid, methyl ester, (Z) (C16:1); hexadecanoic acid, methyl ester (C16:0). This composition is similar to a marine Chlorella sp. in the previous research [17]. The TFA content and the C18 FA (stearic acid, oleic acid, and linoelaidic acid) content of this type of microalgae are low in the favorable cultivation condition [17]. In this research, the C18 FA content was very low, so it could not be detected.

Table 1. The effect of the storage time (one and two days) on total fatty acids (TFAs) and free fatty acids (FFAs) in microalgae.

Fatty acid methyl esters	Microalgae A (mg/g algae)	Microalgae B (mg/g algae)	Microalgae C (mg/g algae)
Myristoleic acid, methyl ester (C14:1)	0.68 ± 0.02	0.88 ± 0.02	0.87 ± 0.04
Tetradecanoic acid, methyl ester (C14:0)	11.73 ± 0.27	13.28 ± 0.27	12.53 ± 0.61
Pentadecanoic acid, methyl ester (C15:0)	0.47 ± 0.01	0.49 ± 0.02	0.50 ± 0.02
9-Hexadecenoic acid, methyl ester, (Z) (C16:1)	9.60 ± 0.19	11.64 ± 0.17	11.97 ± 0.53
Hexadecanoic acid, methyl ester (C16:0)	2.89 ± 0.08	3.00 ± 0.06	3.09 ± 0.15
Total fatty acids (TFAs), M_A, M_B, and M_C	25.37 ± 0.56	29.28 ± 0.49	28.95 ± 1.10
Free fatty acids / Total fatty acids (FFAs/TFAs), %	14.7 ± 0.6	17.7 ± 0.3	18.4 ± 0.3

The percentage of FFAs/TFAs of microalgae A was 14.7%. Inherently, the microalgae have a certain concentration of FFAs. Robert B. Levine and et al., claimed that the FFA content in the dry microalgae (65 °C for 24 h) was 1–2% [6]. On the other hand, Tao Dong and et al. also reported that the FFA content of the lyophilized C. Sorokiniana was very high (46.85% of the TFAs) [22]. During the drying, the lipid hydrolysis by the enzyme lipase was inhibited at a high temperature [23] while the hydrolysis by heat was insignificant at 105 °C [24].

3.1. Effect of the storage time on total fatty acids (TFAs) and free fatty acids (FFAs) in microalgae

According to the data shown in **Table 1**, after the storage period of one day, the amount of TFAs in microalgae increased approximately 15.4% (25.37 to 29.28 mg/g microalgae) compared to the original. This difference is mainly due to the increase of fatty acids (FAs) of C14:0 and C16:1. In the second day, the TFAs gradually reduced a little. Microalgae maybe started to decompose. Meanwhile, there was a slight increase of FFAs/TFAs from 14.7% to 18.4%.

The slight increase of TFAs in microalgae B is consistent with the previous study [12]. The total lipid content of microalgae increased from 33.4% (stored at −80 °C) to 36.6% (stored at 25 °C) [12]. The phenomenon of slightly increased TFA content was also reported by E. Montaini and et al. when the microalgae Tetraselmis suecica paste was stored at 4 °C [25]. On the other hand, the increase of FFAs/TFAs is caused by the hydrolysis of triacylglycerides (TGs) in the presence of enzyme lipase inside the microalgae [12,13]. Due to the small difference between algae B and algae C, the algae B and the original fresh algae A were selected to be used in the subsequent HTC for the comparison.

3.2. Hydrothermal carbonization of microalgae

The mass yields of hydrochar after HTC of microalgae are shown in **Table 2**. The trend is that the higher the HTC temperature was, the less hydrochar could be obtained. However, the difference is negligible. The carbohydrate content of microalgae B is 36.2% which was determined by the colorimetric method [26].

Table 2. Mass yields of hydrochar after HTC of algae.

HTC (°C)	Mass yields of hydrochar, %	
	Algae A	Algae B
200	57.1 ± 0.1	53.6 ± 0.3
210	55.3 ± 0.2	54.7 ± 0.5
220	52.7 ± 0.2	52.7 ± 0.3

Table 3. The TFA retention and FFAs/TFAs in hydrochar after HTC.

HTC (°C)	Total fatty acids (TFAs) retention		$[(M_B \times Y_{B1}) - (M_A \times Y_{A1})] \times 100$ $/(M_A \times Y_{A1})$ %	Percentage of FFAs/TFAs	
	Hydrochar A, Y_{A1}, %	Hydrochar B, Y_{B1}, %		Hydrochar A, %	Hydrochar B, %
200	85.3 ± 1.1	83.2 ± 1.2	12.6	29.4 ± 0.8	37.1 ± 1.2
210	82.9 ± 0.5	78.4 ± 1.3	9.2	31.8 ± 2.7	38.0 ± 0.5
220	73.1 ± 2.4	70.8 ± 2.0	11.9	36.5 ± 0.8	39.3 ± 2.7

The TFA retention in the hydrochar after HTC is presented in **Table 3**. In general, the higher temperature was, the lower TFA retention was due to the enhanced hydrolysis of triacylglycerides (TGs) and the easy extraction of FFAs from microalgae into the aqueous phase under the HTC condition. At 200 °C, the retention level of TFAs was about 85%. This result is consistent with previous studies [2,19,27] which proved that the hydrochar of HTC at around 200 °C retained most of TFAs of original microalgae, from 80% to 95% depending on the type of microalgae.

According to **Table 3**, it can also be seen that hydrochar B had a little bit lower TFA retention compared to hydrochar A ($Y_{B1} < Y_{A1}$) at all HTC conditions. The reason is that compared to microalgae A, microalgae B contained more FFAs which were more freely and easily extracted to water phase during HTC of microalgae due to the high solubility of FFAs in water at a high temperature [14]. Although most of these FFAs were adsorbed again on hydrochar [3], a small part was freely dissolved in the aqueous phase or adsorbed on the dissolved organic matters [15]. The higher the FFA content of microalgae is, the higher the possibility of losing these FFAs in the aqueous phase is. The higher percentages of FFAs/TFAs in hydrochar B also proved that. On the other hand, the amount of TFAs in microalgae B was 15.4% higher than the amount of TFAs in microalgae A ($M_B > M_A$) as mentioned above. Finally, the amount of TFAs in hydrochar B ($M_B \times Y_{B1}$) was still about 10% higher than the amount of TFAs in hydrochar A ($M_A \times Y_{A1}$) at all HTC temperatures. This result means that the additional storage step of microalgae paste can enhance the amount of TFAs in the subsequent hydrochar which favors for biodiesel production.

Meanwhile, the percentage of FFAs/TFAs tends to increase up to 40% as the temperature increases from 200 °C to 220 °C due to enhanced hydrolysis at a higher temperature. This is beneficial for the esterification-transesterification process because FFAs can be converted to FAMEs rapidly under the effect of acid catalysts. However, it should be noted that when the temperature increased up to 220 °C, the TFA retention went down. Therefore, hydrochar A and hydrochar B formed at 200 °C were used in the direct esterification-transesterification process for biodiesel production.

Besides, it can be seen that the percentages of FFAs/TFAs of hydrochar B were higher than those of hydrochar A. This is because the hydrolysis of BFAs occurs slowly at first and then can be accelerated by the self-catalytic effect of formed FFAs [6]. Due to a higher percentage of FFAs/TFAs in microalgae B (17.7% > 14.7%), the hydrolysis rate of BFAs in microalgae B was faster than that in microalgae A at the same HTC condition, and more FFAs were formed in hydrochar B (37.1% > 29.4% at 200 °C). This facilitated the esterification-transesterification using the acid catalyst due to a rapid conversion of FFAs to FAME. It was proved that FFAs can even be converted selectively into FAMEs in the appropriate, mild conditions while BFAs are largely unaffected [28].

Table 4. The elemental analyses for microalgae B and hydrochar B (200 °C).

	Microalgae B	Hydrochar B (200 °C)
C	44.5%	50.9%
H	7.2%	6.9%
N	9.7%	9.2%

The elemental analyses for microalgae B and hydrochar B (200 °C) are shown in **Table 4**. The carbon content increased from 44.5% in microalgae B to 50.9% in hydrochar B (200 °C). The distribution of C and N in two phases is similar to the previous researches [3,5]. For the P distribution, most of P was retained in the hydrochar.

Table 5. The elemental distribution in hydrochar B (200 °C) and the liquid phase.

	Hydrochar B (200 °C)	Liquid phase
C	46.1%	53.9%
N	42.3%	57.7%
P	98.5%	1.5%

The energy balance of the process is calculated based on the HTC of 10 kg of microalgae paste with the moisture content of 85% (8.5 kg of water and 1.5 kg of microalgae) from 25 °C to 200 °C.

- **The energy requirement for HTC of microalgae**

The enthalpies of saturated water are 0.85 MJ/kg at 200 °C (473 K) and 0.10 MJ/kg at 25 °C (298 K) [29], so the heat capacity of water (from 25 °C to 200 °C) is (0.85 MJ/kg − 0.10 MJ/kg) = 0.75 MJ/kg. The heat capacity of microalgae is assumed to be a half of water in this range of temperature [2]. The energy for HTC of microalgae is (0.75 MJ/kg × 8.5 kg) + (0.75 MJ/kg × 0.5 × 1.5 kg) = 6.92 MJ.

- **The energy requirement for drying hydrochar**

The enthalpies of water are 0.42 MJ/kg at 100 °C (373 K) and 0.10 MJ/kg at 25 °C (298 K) [29], so the heat capacity of water (from 25 °C to 100 °C) is (0.42 MJ/kg − 0.10 MJ/kg) = 0.32 MJ/kg. The heat of vaporization of water at 100 °C (373 K) is 2.26 MJ/kg [29]. The mass of hydrochar B (HTC 200 °C) is (1.5 kg × 53.6%) = 0.804 kg. Because the moisture content of hydrochar B after the filtration is around 50%, so the mass of water is 0.804 kg. Finally, the energy requirement for heating and drying hydrochar B is (0.32 MJ/kg + 2.26 MJ/kg) × 0.804 kg = 2.07 MJ. The energy requirement for HTC of microalgae and drying hydrochar is (6.92 MJ + 2.07 MJ) = **8.99 MJ**.

- **Energy balance**

The heating value is calculated based on the Dulong formula HHV (MJ/kg) = 0.338C + 1.428 × (H − O/8) + 0.095 × S [30]. For hydrochar B, the contents of C, H, O and S are 50.9%, 6.9%, 23.3% and 0.5%, respectively, so the heating value of hydrochar B is HHV_1 = 23.0 MJ/kg. The heating value of biodiesel is HHV_2 = 39.9 MJ/kg. Because the fatty acid content in hydrochar B is 0.045 kg/kg hydrochar, the energy balance is calculated as follows.

$$\Delta = (23.0 \text{ MJ/kg} \times 0.804 \text{ kg} - 39.9 \text{ MJ/kg} \times 0.804 \text{ kg} \times 0.045) - 8.99 \text{ MJ} = + \textbf{8.02 MJ}.$$

The energy balance is positive, so the heating value of lipid-extracted hydrochar is higher than the energy requirement for HTC process and drying hydrochar.

3.3. *Direct esterification and transesterification of hydrochar*

Table 6. Yields of the direct esterification-transesterification reaction of hydrochar A (HTC 200 °C) and hydrochar B (HTC 200 °C) at 85 °C, 90min.

Hydrochar (g) /ml methanol	No.	H_2SO_4 conc. % v/v	Hydrochar A Y_{A2}, %	Hydrochar B Y_{B2}, %	$Y_{B2} - Y_{A2}$ %	$(Y_B - Y_A) \times 100/Y_A$, %
0.1	1	1	85.5 ± 0.3	94.4 ± 2.5	8.9	24.4
	2	2	97.2 ± 0.4	98.0 ± 1.0	0.8	13.5
0.2	3	1	63.7 ± 0.8	72.6 ± 1.3	8.9	28.3
	4	2	82.6 ± 0.6	88.7 ± 1.2	6.1	21.1
	5	3	93.7 ± 0.4	99.2 ± 0.3	5.5	19.3
	6	4	96.5 ± 0.5	99.0 ± 0.9	2.5	15.5

$Y_A = M_A \times Y_{A1} \times Y_{A2}$ (mg/g algae), $Y_B = M_B \times Y_{B1} \times Y_{B2}$ (mg/g algae)

The yields of the esterification-transesterification reactions are presented in **Table 6**. Generally, the higher the H_2SO_4 concentration was, the higher the reaction yield was. To obtain the reaction yield exceeding 95%, the required H_2SO_4 concentrations were 2% v/v in the case of 0.1 g char/ml methanol and 4% v/v in the case of 0.2 g char/ml methanol. The methanol solution of 2% v/v H_2SO_4, with the same effect as the methanol solution of 5% v/v HCl, is normally used to prepare fatty acid methyl esters for chromatographic analysis [31]. This H_2SO_4 concentration was still sufficient for the case of 0.1 g char/ml methanol. However, when more hydrochar was used, 0.2 g char/ml methanol, the H_2SO_4 concentration should be higher, 4% v/v, to get the high conversion due to the absorption of H^+ on functional groups such as –COOH, =CO, –OH (mainly) on the surface of hydrochar [32,33]. The absorption capacity was assessed based on the H^+ exchange capacity [33]. The heavy metal cations were also absorbed on these functional groups, and the absorption capacity improved when the number of these functional groups increased [34,35].

Moreover, at the same reaction condition, the reaction yields of hydrochar B were always higher than hydrochar A ($Y_{B2} > Y_{A2}$). This is because the percentage of FFAs/TFAs of hydrochar B is higher than that of hydrochar A (37.1% > 29.4%) as mentioned above, and the acid catalyst H_2SO_4 can catalyze the esterification of FFAs much faster than the transesterification of BFAs [28]. Furthermore, the differences of the reaction yield ($Y_{B2} - Y_{A2}$) being around the difference of the percentage of FFAs/TFAs (37.1% − 29.4% = 7.7%) also prove this explanation, excluding the cases using an amount of sulfuric acid more than necessary for hydrochar B (No.2 and 6).

Finally, the total yields Y_B ($M_B \times Y_{B1} \times Y_{B2}$) and Y_A ($M_A \times Y_{A1} \times Y_{A2}$) are compared as the ratio of ($Y_B - Y_A$) $\times$ 100/Y_A in **Table 6**. Y_B is always around 15%–25% higher than Y_A in all conditions of the esterification and transesterification reactions. For the case of 0.2 g char/ml methanol and 3% v/v H_2SO_4 concentration, the difference is 19.3%. The nitrogen content in biodiesel is 0.33%. In summary, the additional storage step of microalgae not only increased the amount of total fatty acids (TFAs) in the hydrochar product but also made the percentage of FFAs/TFAs higher as well as a higher reaction yield at the same reaction condition. This eventually leads to a higher total yield of the case including the storage of microalgae paste. The comparison of both cases is briefly illustrated in **Figure 4**.

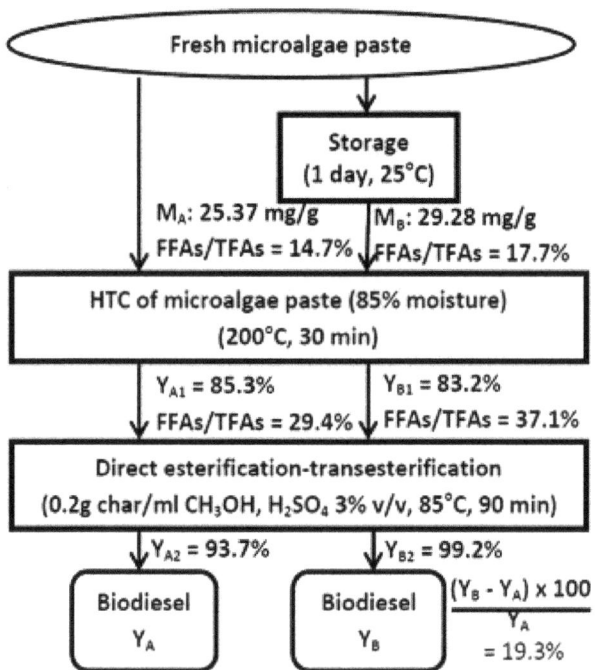

Figure 4. The whole process of biodiesel production from fresh microalgae in both cases of A and B.

4. Conclusion

An entire process, from the storage of fresh algae, the HTC of microalgae paste to the direct esterification-transesterification of hydrochar using the acid catalyst H_2SO_4, was investigated for biodiesel production. Instead of doing HTC of fresh microalgae immediately after harvesting, the storage of fresh microalgae for one day at the room temperature (25 °C) before conducting HTC can improve the total biodiesel yield of 19.3% at the same conversion conditions. This additional step is highly practical because there is no consumption of much energy and chemicals required in the storage of microalgae.

Acknowledgements

The authors thank the financial support from the scholarship of JICA, Japan Cooperation Center International, via Doctoral Degree Program in Japan of AUN/SEED-Net. In addition, the authors give many thanks to Abooali Golzary, Department of Environmental Engineering, Graduate Faculty of Environment, University of Tehran, for his enthusiastic guidance and support in cultivating microalgae.

Conflict of Interest

All authors declare no conflicts of interest in this paper.

References

1. Mata TM, Martins AA, Caetano NS (2010) Microalgae for biodiesel production and other applications: a review. *Renew Sust Energ Rev* 14: 217-232.
2. Heilmann SM, Davis HT, Jader LR, et al. (2010) Hydrothermal carbonization of microalgae. *Biomass Bioenerg* 34: 875-882.
3. Heilmann SM, Jader LR, Harned LA, et al. (2011) Hydrothermal carbonization of microalgae II. Fatty acid, char, and algal nutrient products. *Appl Energ* 88: 3286-3290.
4. Garcia Alba L, Torri C, Samorì C, et al. (2011) Hydrothermal treatment (HTT) of microalgae: evaluation of the process as conversion method in an algae biorefinery concept. *Energ Fuel* 26: 642-657.
5. Levine RB, Sierra COS, Hockstad R, et al. (2013) The use of hydrothermal carbonization to recycle nutrients in algal biofuel production. *Environ Prog Sust Energ* 32: 962-975.
6. Levine RB, Pinnarat T, Savage PE (2010) Biodiesel production from wet algal biomass through in situ lipid hydrolysis and supercritical transesterification. *Energ Fuel* 24: 5235-5243.
7. Folch J, Lees M, Sloane-Stanley G (1957) A simple method for the isolation and purification of total lipids from animal tissues. *J Biol Chem* 226: 497-509.
8. Halim R, Gladman B, Danquah MK, et al. (2011) Oil extraction from microalgae for biodiesel production. *Bioresource Technol* 102: 178-185.
9. Griffiths M, Van Hille R, Harrison S (2010) Selection of direct transesterification as the preferred method for assay of fatty acid content of microalgae. *Lipids* 45: 1053-1060.
10. Tran H-L, Hong S-J, Lee C-G (2009) Evaluation of extraction methods for recovery of fatty acids from Botryococcus braunii LB 572 and Synechocystis sp. PCC 6803. *Biotechnol Bioprocess Eng* 14: 187-192.
11. Lewis T, Nichols PD, McMeekin TA (2000) Evaluation of extraction methods for recovery of fatty acids from lipid-producing microheterotrophs. *J Microbiol Meth* 43: 107-116.

12. Chen L, Liu T, Zhang W, et al. (2012) Biodiesel production from algae oil high in free fatty acids by two-step catalytic conversion. *Bioresource Technol* 111: 208-214.

13. Singh A, Nigam PS, Murphy JD (2011) Mechanism and challenges in commercialisation of algal biofuels. *Bioresource Technol* 102: 26-34.

14. Khuwijitjaru P, Adachi S, Matsuno R (2002) Solubility of saturated fatty acids in water at elevated temperatures. *Biosci Biotechnol Bioch* 66: 1723-1726.

15. Poerschmann J, Weiner B, Wedwitschka H, et al. (2014) Characterization of biocoals and dissolved organic matter phases obtained upon hydrothermal carbonization of brewer's spent grain. *Bioresource Technol* 164: 162-169.

16. Golzary A, Imanian S, Abdoli MA, et al. (2015) A cost-effective strategy for marine microalgae separation by electro-coagulation–flotation process aimed at bio-crude oil production: Optimization and evaluation study. *Sep Purif Technol* 147: 156-165.

17. Sanyano N, Chetpattananondh P, Chongkhong S (2013) Coagulation–flocculation of marine Chlorella sp. for biodiesel production. *Bioresource Technol* 147: 471-476.

18. Broch A, Jena U, Hoekman SK, et al. (2013) Analysis of solid and aqueous phase products from hydrothermal carbonization of whole and lipid-extracted algae. *Energies* 7: 62-79.

19. Lu Y, Levine RB, Savage PE (2014) Fatty Acids for Nutraceuticals and Biofuels from Hydrothermal Carbonization of Microalgae. *Ind Eng Chem Res* 54: 4066-4071.

20. Laurens LM, Quinn M, Van Wychen S, et al. (2012) Accurate and reliable quantification of total microalgal fuel potential as fatty acid methyl esters by in situ transesterification. *Anal Bioanal Chem* 403: 167-178.

21. Thenot J-P, Horning E, Stafford M, et al. (1972) Fatty acid esterification with N, N-dimethylformamide dialkyl acetals for GC analysis. *Anal Lett* 5: 217-223.

22. Dong T, Wang J, Miao C, et al. (2013) Two-step in situ biodiesel production from microalgae with high free fatty acid content. *Bioresource Technol* 136: 8-15.

23. Al-Zuhair S, Hasan M, Ramachandran K (2003) Kinetics of the enzymatic hydrolysis of palm oil by lipase. *Process Biochem* 38: 1155-1163.

24. Khuwijitjaru P, Fujii T, Adachi S, et al. (2004) Kinetics on the hydrolysis of fatty acid esters in subcritical water. *Chem Eng J* 99: 1-4.

25. Montaini E, Zittelli GC, Tredici M, et al. (1995) Long-term preservation of Tetraselmis suecica: influence of storage on viability and fatty acid profile. *Aquaculture* 134: 81-90.

26. Dubois M, Gilles KA, Hamilton JK, et al. (1956) Colorimetric method for determination of sugars and related substances. *Anal Chem* 28: 350-356.

27. Du Z, Mohr M, Ma X, et al. (2012) Hydrothermal pretreatment of microalgae for production of pyrolytic bio-oil with a low nitrogen content. *Bioresource Technol* 120: 13-18.

28. Kail BW, Link DD, Morreale BD (2012) Determination of free fatty acids and triglycerides by gas chromatography using selective esterification reactions. *J Chromatogr Sci* bms093.

29. Perry RH, Green DW (1999) Perry's chemical engineers' handbook: McGraw-Hill Professional.

30. Valdez PJ, Nelson MC, Wang HY, et al. (2012) Hydrothermal liquefaction of Nannochloropsis sp.: Systematic study of process variables and analysis of the product fractions. *Biomass Bioenerg* 46: 317-331.

31. Christie WW (1993) Preparation of ester derivatives of fatty acids for chromatographic analysis. *Adv Lipid Meth* 2: e111.

32. Liu Z, Zhang F-S, Wu J (2010) Characterization and application of chars produced from pinewood pyrolysis and hydrothermal treatment. *Fuel* 89: 510-514.

33. Kang S, Li X, Fan J, et al. (2012) Characterization of hydrochars produced by hydrothermal carbonization of lignin, cellulose, D-xylose, and wood meal. *Ind Eng Chem Res* 51: 9023-9031.

34. Chen Z, Ma L, Li S, et al. (2011) Simple approach to carboxyl-rich materials through low-temperature heat treatment of hydrothermal carbon in air. *Appl Surf Sci* 257: 8686-8691.

35. Liu Z, Zhang F-S (2009) Removal of lead from water using biochars prepared from hydrothermal liquefaction of biomass. *J Hazard Mater* 167: 933-939.

12

Oil extraction from plant seeds for biodiesel production

Yadessa Gonfa Keneni and Jorge Mario Marchetti *

Faculty Sciences and Technology, Norwegian University of Life Sciences, Drøbakveien 31, Ås 1432, Norway

* **Correspondence**: Email: jorge.mario.marchetti@nmbu

Abstract: Energy is basic for development and its demand increases due to rapid population growth, urbanization and improved living standards. Fossil fuels will continue to dominate other sources of energy although it is non-renewable and harm global climate. Problems associated with fossil fuels have driven the search for alternative energy sources of which biodiesel is one option. Biodiesel is renewable, non-toxic, environmental-friendly and an economically feasible options to tackle the depleting fossil fuels and its negative environmental impact. It can be produced from vegetable oils, animal fats, waste oils and algae. However, nowadays, the major feedstocks of biodiesel are edible oils and this has created food vs fuel debate. Therefore, the future prospect is to use non-edible oils, animal fats, waste oils and algae as feedstock for biodiesel. Selection of non-expensive feedstock and the extraction and preparation of oil for biodiesel production is a crucial step due to its relevance on the overall technology. There are three main conventional oil extraction methods: mechanical, chemical/solvent and enzymatic extraction methods. There are also some newly developed oil extraction methods that can be used separately or in combination with the conventional ones, to overcome some disadvantages of the conventional oil extraction methods. This review paper presents, compare and discusses different potential biofuel feedstocks, various oil extraction methods, advantages and disadvantages of different oil extraction methods, and propose future prospective for the improvement of oil extraction methods and sustainability of biodiesel production and utilization.

Keywords: biodiesel; biodiesel feedstock; edible oil; fatty acid; non-edible oil; oil extraction; renewable energy

Abbreviation List

AEOE	Aqueous enzymatic oil extraction
ASE	Accelerated solvent extraction

BP	British petroleum
FAME	Fatty acid methyl esters
FFA	Free fatty acid
IRAR	Infrared radiation assisted reactor
L.	Linnaeus
MAAEE	Microwave-assisted aqueous enzymatic extraction
MAE	Microwave-assisted extraction
PSE	Pressurized solvent extraction
SFE	Supercritical fluid extraction
TAG	Triacylglycerol
UAE	Ultrasound-assisted extraction

1. Introduction

Energy demand is expected to increase due to rapid population growth, expanding urbanization and better living standards [1]. Fossil fuels remain the dominant source of energy [2] though it is non-renewable and has negative impact on global climate [3]. According to BP's Energy Outlook to 2035 [2], world energy consumption is projected to increases by 34% between 2014 and 2035, and fossil fuels remain the dominant source of energy (accounting for almost 80%) powering the global economy in 2035 (down from 86% in 2014). The transport sector, which heavily depends on oil-derived liquid products such as gasoline and diesel, globally occupies the third place when total energy consumption and greenhouse gas (GHG) emissions are considered (after the industry and the building sectors). This consumption level is predicted to increase by 60% by 2030 [4].

Rapid growth in both global energy demand and carbon dioxide emissions associated with the use of fossil fuels has driven the search for alternative energy sources which are renewable and have a lower environmental impact [5,6]. Thus, it has become apparent that biodiesel is destined to make a substantial contribution to the future energy demands of domestic and industrial economies [6]. Biodiesel is produced from vegetable oil or animal fat reacts in the presence of a catalyst (usually a base) with an alcohol (usually methanol) to give the corresponding alkyl esters (for methanol, fatty acid methyl esters) [7]. Biodiesel is non-toxic, biodegradable and a portable fuel produced from renewable sources [3,8] and it is one of the technically and economically feasible options to tackle the fast depletion of fossil fuels and environmental pollution [1]. The other benefit of biodiesel fuel is that it can be used in any mixture with petro diesel fuel, as it has very similar characteristics [3].

The potential feedstocks for biodiesel production are edible (first generation feedstocks) and non-edible vegetable oils (second generation feedstocks), wasted oils and animal fats [9] First-generation biofuels are directly related to a biomass that is generally edible, and are usually produced from edible oils, such as soybeans, palm oil, sunflower, safflower ,rapeseed, coconut and peanut [4,10]. Second-generation biofuels are fuels that are produced from a wide array of different feedstock, ranging from lignocellulosic feedstocks to municipal solid wastes. Third-generation biofuels are related to algae which have been considered as emerging non-edible oil sources of growing interest because of their high oil content and rapid biomass production [10,11,12] but could also to a certain extent be linked to utilization of CO_2 as feedstock [10]. However, the first generation biofuels seems to create some skepticism to scientists. There are concerns about environmental impacts and carbon balances, which sets limits in the increasing production of biofuels of first generation. The main disadvantage of first generation biofuels is the food-versus-fuel debate, one of the reasons for rising food prices is due to the increase in the production of these fuels [9,13,14]. Therefore, non-edible biodiesels feedstocks get great attention to overcome the problem that occurs due to continuous utilization of edible oils for biodiesel [13].

In the different literature, various biodiesel feedstocks: edible oils, non-edible oils, animal fats, waste oils and algal biomass and methods of biodiesel production from these feedstocks were well described and reviewed. However, the preparation of different feedstocks for oil extraction, oil extraction methods from different feedstocks, advantages and disadvantages of the extraction methods and ways to improve them are, to our knowledge, not yet well reviewed. Thus, the aim of this review is to identify the major biodiesel feedstocks, oil extraction and separation methods, the advantages and disadvantages of various oil extraction methods, particularly that of non-edible oils, and suggest how to optimize the appropriate method(s) to enhance the sustainability of biodiesel production and utilization.

2.   Biodiesel and Its Feedstock

Biodiesel is defined as the mono-alkyl ester of long chain fatty acids derived from renewable lipid feedstock such as vegetable oils or animal fats [15]. Biodiesel is a non-toxic, biodegradable and renewable fuel that can be produced from a range of organic feedstock including fresh or waste vegetable oils, animal fats, and oilseed plants [16] (the reaction for biodiesel formation is shown in Figure 1).

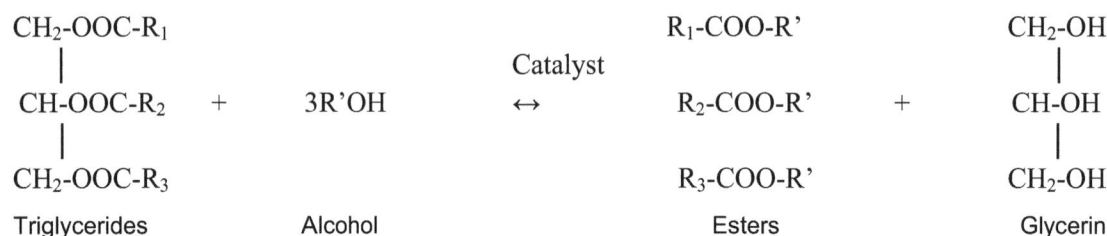

Figure 1. Transesterification reaction for biodiesel production [17,18].

The major components of plant oils and animal fats are triacylglycerol (TAGs); the esters of fatty acids and glycerol. The TAGs, also known as triglycerides, consists of different fatty acid composition which influences both physical and chemical properties of plant oils and animal fats [15,18,19]. There are two kinds of fatty acids: saturated fatty acids containing carbon-carbon single bond, and unsaturated fatty acids which include one or more carbon-carbon double bond. The major components of biodiesel are straight fatty acid chain and the common fatty acids are palmitic acid (C16:0), stearic acid (C18:0), oleic acid (C18:1), linoleic acid (C18:2) and linolenic acid (C18:3). The other fatty acids which are also present in several plant oils include myristic acid (C14:0), palmitoleic acid (C16:1), arachidic acid (C20:0), and erucic acid (C22:1) [11,18,19]. According to Sajjadi et al. [17], physico-chemical properties of oils from different sources differ, and it is noteworthy that although some oils may be extracted from a unique root, their compositions are significantly dependent on the main parts from which the oil is extracted.

2.1. Different types of oils

Globally, there are more than 350 oil-bearing crops identified as potential sources for biodiesel production [13,17]. The availability of wide range of biodiesel feedstocks is one of the most significant factors that enables the sustainable production of biodiesel [20,21,22]. According to Avhad and Marchetti [18], satisfactory replacement of petroleum diesel with biodiesel depends on two basic requirements: first is its easy availability and environmentally acceptability, and the second is being economically reasonable. Availability of feedstock for producing biodiesel depends on the regional climate, geographical locations, local soil conditions and agricultural practices of any country [13].

From the literature, it has been found that feedstock alone represents about 75% of the overall biodiesel production cost [13,23,24] as presented in Figure 2. Therefore, minimizing the cost of biodiesel production has been the main agenda for biodiesel producers in order to be competitive with petroleum-derived diesel [25]. Hence, it is crucial to employ inexpensive feedstocks to replace expensive refined oils [4,13]. Using low-cost triglyceride sources such as waste cooking oil and animal fats is also important to minimizing the total cost as these wastes are three times cheaper than refined oils, and are abundantly available [25].

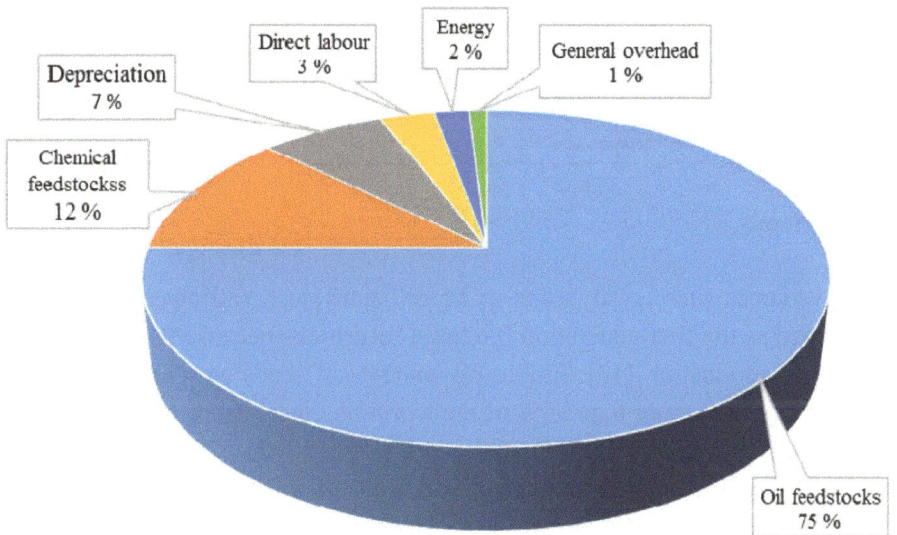

Figure 2. General cost breakdown for production of biodiesel [13,23,24].

Feedstocks of biodiesel can be divided into four main categories: edible vegetable oil, non-edible oils, waste or recycled oils, and animal fats [13,18,23,26,27,28]. Accordingly, some forms of plant oils, animal fats, and other feedstocks that are used for producing biodiesel are listed in Table 1.

Table 1. Main feedstocks of biodiesel [18,20,29–36].

Edible oils	Non-edible oils	Animal fats	Other sources
Barley	*Abutilon muticum*	Beef tallow	Cyanobacteria
Canola	*Aleurites moluccana*	Chicken fat	Bacteria
Coconut	Camelina (*Camelina Sativa*)	Fish oil	Cooking oil
Corn	Coffee ground (*Coffea arabica*)	Pork lard	Fungi
Groundnut	Cotton seed (*Gossypium hirsutum*)	Poultry fat	Latexes
Palm and palm kernel (*Elaeis* guineensis)	*Croton megalocarpus*	Waste salmon	Microalgae (Chlorellavulgaris)
Peanut	*Cynara cardunculus*		Miscanthus
Pumpkin seed	*Jatropha curcas*		Pomace oil
Rapeseed (*Brassica napus L.*)	Jojoba (*Simmondsia chinensis*)		Poplar
Rice bran oil (*Oryza sativum*)	Karanja or honge (*Pongamia pinnata*)		Soapstocks
Safflower(*Carthamus tinctorius*)	Mahua (*Madhuca indica*)		Switchgrass
Sesame (*Sesamum indicum L.*)	Moringa (*Moringa oleifera*)		Tall oil
Sorghum	Nagchampa (*Calophyllum inophyllum)*		Tarpenes
Soybeans (*Glycine max*)	Neem (*Azadirachta indica*)		
Sunflower (Helianthus annuus)	*Pachira glabra*		
Wheat	Passion seed (*Passiflora edulis*)		
	Pongamia (*Pongamia pinnata*)		
	Rubber seed tree (*Hevca brasiliensis)*		
	Terminalia belerica		
	Tobacco seed		

2.1.1.   Edible plant oils

Edible oils resources such as soybeans, palm oil, sunflower, safflower, rapeseed, coconut and peanut are considered as the first generation biodiesel feedstocks because they were the first crops to be used for biodiesel production [13]. Edible oil contents of seeds and kernels of some plants are shown in Table 2. Currently, more than 95% of the world biodiesel is produced from edible oils such as rapeseed (84%), sunflower oil (13%), palm oil (1%), soybean oil and others (2%) [4,13]. Plantations of these feedstock plants have been also well established in many countries around the world such as Malaysia, USA and Germany [13]. However, continuous large-scale usage of edible plant oils for biodiesel production raises many concerns such as food versus fuel crisis and major environmental problems such as deforestation and destruction of vital soil resources, conversion of much available farm lands to oil bearing plants [13,14].

The prices of vegetable oil have also increased dramatically in the last few decades and this will affect the economic viability of biodiesel industry [13,14,51]. Furthermore, the use of such edible oils to produce biodiesel is not feasible in the long term due to the growing gap between demand and supply [13]. Thus, the current use of the food-grade plant oils as a feedstock for producing biodiesel are considered to be not worthy and stipulates search for relatively less expensive resources [13,18].

Table 2. Species name and oil content of edible and non-edible plant [6,11,12,13,30,37–50].

Type of oil	Common name	Species name	Oil content of seed/kernel (wt%)		Reference
			Seed	Kernel	
Edible	Coconut	*Cocos Nucifera L.*	63–65	63.1 (± 2.8)	[6,38]
	Corn	*Zea mays*	24.44	-	[39]
	Hemp seed	*Cannabis Sativa L.*	22–38	-	[40]
	Mustard seed	*Brassica nigra*	33	-	[41]
	Olive	*Olea europaea*	45–70	-	[37]
	Palm	*Elaeis guineensis*	30–60	-	[6]
	Peanut	*Arachis hypogea L.*	45–55	47–61	[6,37]
	Pumpkin seed	*Cucurbita maxima*	31.5	43.69 (± 3.92)	[42,43]
	Rapeseed	*Brassica napus*	38–46	-	[6]
	Rice bran	*Oryza sativa*	15–23	-	[37]
	Safflower seed	*Carthamus tinctorius*	35	-	[44]
	Sesame seed	*Sesamum indicum*	58	-	[45]
	Soybean	*Glycine max*	18–20	-	[46]
	Sunflower	*Helianthus annuus*	25–35	50	[37,47]
Non-edible	Castor	*Ricinus communis L.*	45–50	-	[30]
	Cottonseed	*Gossypium hirsutum L.*	18–25	31.42	[37,48]
	Desert date	*Balanites aegyptiaca*	45–50	36–47	[12]
	Jatropha	*Jatropha curcas L.*	20–60	40–60	[12,30,49]
	Jojoba	*Simmondsia chincnsis*	45–50	-	[37]
	Karanja	*Pongamia pinnata*	30–40	30–50	[30]
	Linseed	*Linum usitatissimum*	35–45	-	[30]
	Mahua	*Madhuca indica*	35–40	50	[30]
	Neem	*Azadirachta indica*	20–30	25–45	[30]
	Polanga	*Calophyllum inophyllum*	65	22	[12]
	Caster	*Ricinus communis*	45–50	-	[11]
	Rubber seed	*Hevea brasiliensis*	40–60	40–50	[12]
	Tobacco	*Nicotiana tabacum L.*	30–43	-	[37]
		Nicotiana tabacum	36–41	17	[11,12]
		Zanthoxylum bungeanum	24–28	25	[12]
	Tung	*Vernicia montana*	16–18	-	[37]
	Ethiopian mustard	*Brassica carinata*	42	2.2–10.8	[12]
	Sea mango	*Cerbera odollam*	54	6.4	[12]
	Croton oil plant	*Croton tiglium*	30–45	50–60	[50]

The average fatty acid composition of different edible vegetable oils are shown in Table 3. The dominant fatty acids of edible oils are oleic acid (C18:1), linoleic acid (C18:2), palmitic acid (C16:0) and staeric acid (C18:0). The fatty acid composition of edible oils from different plants seeds differ. For example, caprylic acid, which is the lightest compound is only available in wheat grain (11.4 wt%) and coconut oils (8.45 wt%) [17]. According to Sajjadi et al. [17], generally is assumed that the compositions of fatty acids compositional profiles do not change during the conversion of the feedstocks to fuel via transesterification and thus, greatly affect the quality of biodiesel to be produced.

Table 3. Comparison of the fatty acid composition of the selected edible oils.

Source	Fatty Acids Composition													Reference
	C14:0 Myristic acid	C16:0 Palmitic acid	C16:1 Palmitoleic acid	C18:0 Stearic acid	C18:1 Oleic acid	C18:2 Linoleic acid	C18:3 Linolenic acid	C20:0 Arachidic acid	C22:0 Behenic acid	C20:1 Gadoleic acid	C22:1 Erucic acid	C24:0 Lignoceric acid	C8:0 Caprylic acid	
Waste coconut oil	0.50	21.40	0.20	3.00	27.50	47.40	-	-	-	-	-	-	8.45	[52]
Corn	-	-	11.67	1.85	25.16	60.60	0.48	0.24	-	-	-	-	-	[15]
Hempseed	-	6.0–8.5	-	2.5–3.0	12.0–15.0	52.0–56.0	-	0.5–0.8	-	-	-	-	-	[53]
Mustard seed	0.05	5.54	0.21	1.51	8.83	10.79	20.98	1.21	1.09	5.27	37.71	1.68	-	[41]
Olive	-	11.60	1.00	3.10	75.00	7.80	0.60	0.30	0.10	-	-	0.50	-	[54]
Palm	0.70	36.70	0.10	6.60	46.10	8.60	0.30	0.40	0.10	0.20	-	0.10	-	[54]
Peanut	0.20	8.0	-	1.80	53.30	28.40	0.30	0.90	3.00	2.40	-	1.80	-	[54]
Pumpkin seed	-	13.80	-	11.20	29.50	45.5	-	-	-	-	-	-	-	[42]
Rapeseed	-	-	3.49	0.85	64.40	22.30	8.23	-	-	-	-	-	-	[15]
Rice bran	-	22.00	-	3.00	38.00	35.00	-	-	-	-	-	-	-	[55]
Safflower seed	-	11.07 (± 0.10)	-	4.37 (± 0.10)	12.76 (± 0.22)	69.65 (± 0.24)	0.49 (± 1.15)	0.78 (± 0.05)	0.59 (± 0.09)	-	-	0.29	-	[56]
Sesame	-	9.80 (± 0.21)	-	6.3 (± 0.15)	41.82 (± 0.91)	40.50 (± 1.01)	0.32 (± 0.01)	0.67 (± 0.03)	-	-	-	(± 0.13)	-	[57]
Sunflower	-	16.29 (± 0.54)	-	6.66	22.70 (± 0.07)	44.13 (± 0.60)	8.97 (± 0.52)	0.62 (± 70.11)	0.63 (± 70.02)	-	-	-	-	[55]
Soybean	-	6.14	0.09	4.11	34.30	51.17	2.23	0.17	0.41	-	0.53	-	-	[58]
Wheat grain	0.13	17.71	0.2	0.78	16.5	56	2.9	-	-	-	-	-	11.4	[17]

One of the possible solutions to reduce the utilization of the edible oil for biodiesel production is by exploiting non-edible oils. They got great attention as the plants from which these oils obtained are easily available in many parts of the world [6,13,59]. These plants can grow on wastelands that are not suitable for food crops, eliminate competition for food, reduce deforestation rate, and their oils are very economical compared to edible oils [13].

2.1.2. Non-edible plant oils

Non-edible plant oils which are known as the second generation feedstocks can be considered as promising substitutions for traditional edible food crops for the production of biodiesel [6]. Recently, these oils have gained enormous attention as a new generation feedstock because of their high oil content, easy availability, and having the advantage that it could be grown on lands which are not suitable for agriculture [6,13]. Non-edible oil bearing plants could also be grown with less intensive attention; thus, reducing the cost of cultivation [6,12,13,18]. Therefore, production of biodiesel from non-edible oils is an effective way to overcome the associated problems with edible oils [6]. Some of the main and most investigated non-edible plant oils for biodiesel production include jatropha seed oil [32,36], karanja oil [33], jojoba oil [34], linseed oil [35], cottonseed oil [60], amongst others (Table 1).

During selection of any feedstock as a biodiesel source, the amount of oil that can be obtained from the seeds or kernel is an important parameter. Estimated oil contents of seed and kernel of some non-edible vegetable oil were shown in Table 2 [6,12]. Moreover, fatty acid composition is an important characteristic of biodiesel feedstock as it determines the efficiency of process to produce biodiesel. It has been observed that the percentage and type of fatty acid compositions depends mainly on the plant species as well as their growth conditions [6].

The fatty acid composition and distribution of non-edible oils are generally aliphatic compounds with a carboxyl group at the end of a straight chain [4]. Ong et al. [19] reported that the presence of fatty acid compositions has interfered fuel properties and quality of biodiesel. It has also been found that the biodiesel with a high level of methyl oleate (mono unsaturated fatty acid) might have excellent characteristics in ignition quality, fuel stability and flow properties at low temperature [19,61].

Generally, non-edible oil is composed of a high number of double carbon chain (polyunsaturated acid) which indicate that the these oil group has a greater degree of unsaturated fatty acid than saturated carbon chain [19,62]. Moreover, it was reported that cetane number, heat of combustion, melting point, and viscosity of neat fatty compounds increase with increasing chain length and decrease with increasing unsaturation [62,63] of the fatty acid methyl esters (FAME) molecule. Therefore, structural fatty acid composition will affect the physico-chemical properties of biodiesel such as cetane number, cold flow properties, heat of combustion and viscosity [6,61,62]. Fatty acid compositions of various non-edible oils that were found to be suitable for production of biodiesel are shown in Table 4.

Table 4. Comparison of the fatty acid composition of the selected non-edible plant oils [4,17,30,32,34,37,64–70].

| Feed stocks | Fatty Acids | | | | | | | | | | | | Reference |
	C14:0 Myristic acid	C16:0 Palmitic acid	C16:1 Palmitoleic acid	C18:0 Stearic acid	C18:1 Oleic acid	C18:2 Linoleic acid	18:3 Linolenic acid	C20:0 Arachidic acid	C22:0 Behenic acid	C20:1 Gondoic acid	C22:1 Erucic acid	C18:1 Riconoleic acid	
Caster seed	-	1.00	-	-	3.00	5.00	1.00	-	-	-	-	89.00	[64]
Cottonseed	1.00	25.80	0.60	2.5	16.4 (± 0.8)	51.50	0.20	0.20	0.20	-	-	-	[65]
Desertdate kernel	-	15.40 (± 0.26)	-	19.01 (± 0.29)	25.74 (± 0.35)	39.85 (± 0.48)	-	-	-	-	-	-	[66]
Jatropha	-	15.20	0.70	6.80	44.60	32.20	-	0.40	-	-	-	-	[32,34]
Jojoba	-	1.20	-	-	10.70	-	-	9.10	-	59.50	12.30	-	[67]
Karanja	-	11.65	-	2.4–8.9	51.59	16.46	2.65	-	-	-	-	-	[68]
Linseed	-	5.10	0.30	2.5	18.90	18.10	55.10	-	-	-	-	-	[69]
Mahua	-	17.80	-	14.00	46.30	17.90	-	-	-	1.7	-	-	[70]
Moringa	-	7.60	1.40	5.5	66.60	8.10	0.20	5.80	-	-	-	-	[71]
Neem	0.2–0.26	14.9	0.1	20.6	43.9	17.9	0.4	1.6	0.3	-	-	-	[72]
Polonga	-	12.01	-	12.95	34.09	38.26	0.30	-	-	-	-	-	[68]
Rubber seed	2.2	10.2	-	8.7	24.6	39.6	16.3	-	-	-	-	-	[73]
Tobaco	0.14	8.46	-	3.38	11.24	75.58	1.14	-	-	-	-	-	[74]
Tung	-	4.00	-	1.00	8.00	4.00	3.00	-	-	-	-	-	[37]

2.1.3. Waste oils and animal fats

The residual obtained after using oil for the cooking purposes is generally discarded with no further application [18]. Over the last few years, waste cooking oil has been considered as a possible feedstock for biodiesel production due its low cost, and as its biofuel was found to fulfill the requirements specified by European standard for biodiesel (EN) and American Society for Testing and Materials (ASTM) standards [75]. However, waste oil is highly impure consisting mainly of high free fatty acid (FFA), and thus, could be categorized in two groups based on its FFA content: the yellow grease (FFA < 15%), and the brown grease (FFA > 15%). These oils after the filtration and purification processes could be used for biodiesel production [18].

Animal fats such as tallow [76], chicken fat [77], lard [78] and yellow grease [79] are also considered as feedstocks. According to Adewale et al. [80], animal fat wastes are low cost, mitigate environmental damage and increase the quality of the resultant biodiesel fuel. However, it has been reported that these may not be plentiful enough to satisfy the global energy demand. Moreover, biodiesel derived from animal fats has a relatively poor performance in cold weather. Furthermore, the transesterification process is difficult for some types of fats due to the presence of a high amount of saturated fatty acids. [4,13].

2.1.4. Algae as biodiesel feedstocks

The amounts of oily crops, both edible and non-edible, animal fats and waste cooking oils are limited, so it is unlikely to provide worldwide biodiesel production demand. The search for other renewable sources is needed to provide the required amount of oily feedstocks. In recent years a high interest towards producing biodiesel from microalgae has been developed. The advantages of using microalgae for biodiesel production are: much higher biomass productivities than land plants, some species can accumulate up to 20–50% triacylglycerol, no agricultural land is required to grow the biomass and they required only sunlight and a few simple and cheap nutrients [81].

3. Oil Extraction Methods

One of the important steps in the production of biodiesel is oil extraction, and different methods and techniques of oil extraction are in use [4,12,13]. Preparation of feedstocks and various oil extraction methods are discussed in the following parts.

3.1. Feedstock preparation

The pre-requisite for oil extraction is seed preparation [4,13,82]. The preparation of seeds involves removal of outer layers of the fruit to expose the kernels or seeds, and its drying to reduce moisture content [82]. The seeds are separated from fruits, and the fruits that do not dehisce are cracked open manually. The separated seeds or kernels are sieved, cleaned and stored at room temperature [13].

According to Atabani et al. [13,82] seeds can be either dried in the oven or sun dried to appropriate moisture. The kernels or seeds have to be prepared in such a way that they contain optimum moisture content for high oil extraction. For instance, Jahirul et al. [82] has found that seed kernel of beauty leaf (*C. inophyllum*) prepared to 15% moisture content provided the highest oil yields in both mechanical and solvent extraction methods. The drying process should be checked very carefully by weighing the trays several times in a day whenever possible and after reaching the desired dryness; the trays are stored in a refrigerated room [4]. Mechanical expellers or presses can be fed with either whole seeds or kernels or a mix of both, but common practice is to use seeds only. However, for chemical extraction only kernels are employed [83].

3.2. Extraction methods

After preparation, the raw material is ready for oil extraction. There are three main methods that have been identified for oil extraction: (i) mechanical extraction, (ii) chemical or solvent extraction, and (iii) enzymatic extraction [4,6,13]. Moreover, accelerated solvent extraction (ASE), supercritical fluid extraction (SFE) as well as microwave-assisted extraction (MAE) methods are frequently used [4]; however, they are not as common or well known as the first three mentioned alternatives

It has been observed that mechanical pressing and solvent extraction are the most commonly used methods for commercial oil extraction [6]. According to Atabani et al. [13], the main products during oil extraction are the crude oil, and the important by-products are such as seeds or kernel cakes. Seed cakes can be used as fertilizers for soil enrichment [6], feed for poultry, fish and swine, and some oil cakes have also application in fermentation and biotechnological processes [84].

3.2.1.   Mechanical oil extraction

Mechanical press oil extraction is the most conventional technique. A manual ram press or an engine driven screw press can be used [4]. Jahirul et al. [82] and Bhuiya et al. [85] used a Mini 40 screw press to extract oil from beauty leaf kernels (*C. inophyllum*). It has been found that engine driven screw press can extract 68–80% of the available oil while the ram presses only achieved 60–65%. Oil extraction efficiencies calculated from data reported in more recent studies are found to generally correspond to these ranges, although the efficiency range of engine driven screw presses can be broadened to 70–80% [4,6,13]. This broader difference is due to the fact that seeds can be subjected to a different number of extractions through the expeller [82,85]. Calculated oil yields (% of contained oil) of mechanical extraction method is presented in Table 5.

Table 5. Calculated oil yields (% of contained oil) of mechanical extraction methods [6,82,83,86].

Press type	Oil yield (%)	Necessary treatment
	68.0	
Engine driven screw press	80.0	Filtration and degumming
	79.0	
Ram press	62.5	

The oil extracted by mechanical presses needs further treatment of filtration and degumming in order to produce a more pure raw material [6,87]. Another problem associated with conventional mechanical presses is that the design of mechanical extractor is suited for some seeds, and therefore, the oil yield is affected if that mechanical extractor is used for other seeds [4,6,13,87]. It has been also found that pretreatment of seeds before applying mechanical extractor increases the amount of oil recovery [6,83]. For instance, by cooking jatropha seeds in water for one hour at 70 °C and using screw pressing, Beerens [88] obtained oil yield of 89% after single pass and 91% after dual pass compared to 79% and 87% oil yield recovery of untreated seeds, respectively. Therefore, several other methods have been proposed recently for oil extraction such as solvent extraction, enzymatic extraction and microwave assisted techniques in order to improve the oil extraction yield.

3.2.2. Solvent oil extraction (chemical extraction)

Solvent extraction is the process in which the oil is removed from a solid by means of a liquid solvent, it is also known as leaching [4]. The chemical extraction using n-hexane method results in the highest oil yield which makes it the most commonly used solvent [4,13]. Jahirul et al. [82] has used n-hexane to extract the oil from Australian native beauty leaf seeds (*Calophyllum inophyllum*), although the cost of oil extraction technique by mechanical screw press is low it is ineffective due to relatively lower oil yields. On the contrary, the chemical oil extraction method was found to be very effective because of high oil yield and for its consistent performance.

Table 6. Advantages and disadvantages of mechanical and chemical oil extraction methods for beauty leaf seeds [82].

Mechanical Extraction		Chemical Extraction	
Advantages	**Disadvantages**	**Advantages**	**Disadvantages**
✓ Virgin oil is more sought after	• Generally ineffective for processing Beauty Leaf seed	✓ Repeatable and reproducible results and process	• Less sought after than virgin oil
✓ No potential for solvent contamination	• Time and labor intensive	✓ High oil yields	• Potential for solvent contamination
✓ Relatively inexpensive after initial capital costs	• Relatively low oil yields	✓ Relatively simple and quick	• Safety issues and environmental concerns regarding the use of hexane
✓ Minor consumables cost	• Operators require experience to achieve best results	✓ Hexane can be recovered and reused, reducing cost significantly	• Very costly if the hexane cannot be recovered
	• High dependence on kernel moisture content		

It has been observed that there are many factors affecting the rate of solvent extraction such as particle size, the type of solvent used, temperature and agitation speed [6,13]. The solvent has to be selected in such a way that it would be a good selective solvent and its viscosity would be sufficiently low to circulate freely. Sayyar et al. [89] extracted J. curcas oil by n-hexane and petroleum ether and found that the extraction yield with n-hexane to be about 1.3% more than that of petroleum ether (47.3% and 46.0% wt, respectively) under similar conditions. The authors recognized n-hexane as a more preferable solvent for extraction of jatropha oil as compared to petroleum ether. In the extraction of olive oil using organic solvents like hexane, ethanol, petroleum ether, isopropyl alcohol and carbon tetrachloride by a Soxhlet extractor, Banat et al. [90] did also obtain the highest oil yield (12.7%) by n-hexane. However, it has been observed that this method consumes much more time compared to other techniques. The solvent extraction is only economical attractive at a large-scale of production (more than 50 ton biodiesel per day) as reported [13]. In addition, n-hexane solvent extraction has a negative environmental impact because of the wastewater generation, higher specific energy consumption and higher emissions of volatile organic compounds and human health impacts [6]. According to Achten et al. [83] and Mahanta and Shrivastava [91], there are three other types of solvent extraction technique: hot water extraction, soxhlet extraction and ultrasonication technique that could be use instead of hexane solvent extraction.

Jahirul et al. [82] reported that in oil extraction from beauty leaf seeds *(Calophyllum inophyllum)* by mechanical method (using the screw press) and chemical extraction (using hexane as a solvent), each method has advantages and disadvantages. The advantages and disadvantages of oil extraction by mechanical extraction and chemical extraction from beauty leaf seeds is presented in Table 6.

3.2.3. Accelerated solvent extraction (ASE)

Accelerated solvent extraction (ASE) is also referred to as pressurized solvent extraction (PSE) is another modern oil extraction technique which uses organic and/or aqueous solvents at elevated temperatures and pressures [4]. It has been observed that high temperature accelerates the extraction rate, while elevated pressure prevents boiling at temperatures above the normal boiling point of the solvent.

Khattab and Zeitoun [92] have extracted oil of flaxseed by different methods by supercritical fluid extraction (SFE), accelerated solvent extraction (ASE) and conventional solvent extraction (SE) and found the highest oil yield (42.40%) by SE using n-hexane which did not differ significantly from the one obtained by accelerated solvent extraction ASE in terms of oil quantity (41.90%) and their physicochemical properties and fatty acid profiles. The supercritical fluid extraction (SFE), however, showed significantly lower oil yield (36.49%) in this particular oil extraction from flaxseed. Sarip et al. [93] have also extracted crude palm oil from palm mesocarp by using hot compressed water extraction method and obtained $70 \pm 0.5\%$ of the oil with averaged free fatty acid of $0.81 \pm 0.08\%$. Moreover, it was also reported that ASE has been used for the extraction of different materials including wheat germ [94] and flaxseed hulls [95]. In ASE, time as well as solvent consumption is significantly reduced compared to the other solvent extraction techniques [92,94].

3.2.4. Enzymatic oil extraction

Aqueous enzymatic oil extraction (AEOE) method is a promising technique for extraction of oil from plant materials [96,97]. In this method, enzymes should be used to extract oil from crushed seeds [91]. Aqueous enzymatic oil extraction can also be used in combination with other methods of oil extraction. For instance, Shah et al. [97] used a combination of ultrasonication and aqueous enzymatic oil extraction (using an alkaline protease at pH = 9.0) method to extract oil from *J. curcas* seeds and obtained 74% of the seed oil which is very large compared to the 17–20% oil extracted by aqueous oil extraction alone. Moreover, using of ultrasonication also resulted in reducing the process time from 18 to 6 h. The main advantages of using enzymatic oil extraction are that it is environmental-friendly and does not produce volatile organic compounds. However, the long process time is the main disadvantage associated with this technique [91].

Table 7 shows the reaction temperature, reaction pH, time consumption and oil yield of different chemical and enzymatic extraction methods tested on *J. curcas*. It has been found that the chemical extraction using n-hexane method results in the highest oil yield which makes it the most commonly used method. Moreover, the negative environmental impacts associated with solvent extraction can be reduced significantly by using AEOE technique although the later method takes long time to complete the process [83,91].

Table 7. Reported oil yields percentage for different chemical and enzymatic extraction methods and different reaction parameters for *J. curcas*.

Extraction technique	Reaction temperature (°C)	Reaction pH	Time consumption (h)	Oil yield (%)	Reference
n-Hexane oil extraction (Soxhlet) apparatus	-	-	24	95–99	[86]
First acetone, second n-hexane	-	-	48	-	[98]
AOE	50	9	6	38	[83,97]
AOE with 10 min of ultrasonication as pre-treatment	50	9	6	67	[83,97]
AEOE (hemicellulase or cellulase)	60	4.5	2	73	[6,13,83]
AEOE (alkaline protease)	60	7	2	86	[6,13,83]
	50	9	6	64	[83]
AEOE (alkaline protease) with 5 min of ultrasonication as pre-treatment	50	9	6	74	[83,97]
Three-phase partitioning	25	9	2	97	[83,99]

3.2.5. Supercritical fluid extraction (SFE)

Supercritical fluid extraction (SFE) technique is used to avoid the use of organic solvents and to increase the speed of extraction [4]. SFE using CO_2 has numerous advantages over the solvent extraction [92,100]. It uses CO_2 as a solvent which is a nontoxic, inexpensive, nonflammable, and nonpolluting supercritical fluid solvent for the extraction of natural products, and also almost 100% oil can be extracted by this method [100].

Maran and Priya [101] have used a supercritical fluid extraction (at 44 MPa, 49.8 °C, and 0.64 g/min of CO_2 flow rate and within 81 min) method for extraction of oil from muskmelon seed (*Cucumis melo*) and produced slightly higher oil yield (48.11 ± 0.04%) than that of Soxhlet extraction method (46.83 ± 0.29%). Moreover, these authors reported that the fatty acids composition of muskmelon seed oil extracted by SFE was similar to that of Soxhlet extraction. However, the main limitation of the SFE is the high cost at production scale, not only due to the use of high pressure equipment but also because of the raw material should be freeze dried to reduce its moisture to values below 20%, as high water concentration in fluid phase negatively affects the oil yield [102,103].

3.2.6. Microwave-assisted extraction (MAE)

Microwave-assisted extraction (MAE) also called microwave extraction, is a new extraction technique, which combines microwave and traditional solvent extraction [104]. MAE has been recognized as a technique with several advantages over other extraction processes, such as reduction of costs, shorter time, less solvent, higher extraction rate, better products with lower cost, reduce energy consumption and CO_2 emissions [104,105]. In microwave-assisted aqueous enzymatic extraction (MAAEE) of pumpkin seed oil by using mixtures of cellulose, pectinase and proteinase (w/w/w), Jiao et al. [106] obtained the highest oil recovery of 64.17%. The authors also reported that there were no significant variations in physicochemical properties of MAAEE and soxhlet extracted oils, and thus, MAAEE is a promising and environmental-friendly technique for pumpkin seed oil extraction. Moreover, it has been found that the MAE method needs a few minutes compared to SFE and the apparatus of MAE extraction is simpler and cheaper, and can be used with a variety of materials with less limit of the polarity of extractants [104]. Therefore, MAE extraction is an interesting alternative to conventional liquid solvent extraction methods, especially in the case of plant material [4,104]. In microwave-assisted solvent extraction of oil from soybeans and rice bran by using solvent (ethanol) to feedstock ratio of 3:1, the maximum oil yields of 17.3% and 17.2% at 20 min and 120 °C were achieved as compared to 11.3% and 12.4% using control extraction for soybeans and rice bran, respectively [107].

4.   Advantages and Disadvantages of Main Oil Extraction Methods

From the above discussions, it is possible to observe that each method of oil extraction has its own advantages and disadvantages. The advantages and disadvantages of the main three oil extraction methods: mechanical, chemical or solvent and ASE are summarized in Table 8.

Table 8. Advantages and disadvantages of main three oil extraction methods [4,82,108].

Methods	Advantages	Disadvantages
Oil press	✓ Virgin oil is more sought after	• Generally ineffective in beauty leaf oil extraction
	✓ No potential for solvent contamination	• Time and labor intensive
	✓ Relatively inexpensive after initial capital costs	• Relatively low oil yields
	✓ Minor consumable costs	• Operators require experience to achieve best results
	✓ Whole seeds or kernels can be processed	• High dependence on kernel moisture content
	✓ No environmental problem regarding the use of screw press	• Relatively dirty process
		• Filtration or degumming process pf oil is required
		• Low and inconsistent oil production
		• High oil loss
n-Hexane	✓ Repeatable and reproducible results and process	• Less sought after than virgin oil
	✓ High oil yields	• High potential for solvent contamination
	✓ Relatively simple and quick	• Safety issues and environmental concerns
	✓ Suitable for bulk oil extraction	• Very costly if the hexane cannot be recovered
	✓ Low capital investment	• High hexane requirement
	✓ No especial equipment required	• Only kernel can be processed
	✓ Hexane can be recovered and reused, reducing cost significantly	
ASE	✓ Automatic technique	• Very high initial cost
	✓ Condition can be optimized	• High preparation required
	✓ More efficient	• Special equipment and skill required
	✓ Clean process	• Potential for solvent contamination
	✓ Relatively less solvent consumption	• Only kernel can be processed
	✓ Less time and labor incentives	
	✓ High oil yield	

5. Single and Combined Oil Extraction Methods to Reduces Problems of Extraction

Traditional oil extraction methods have their own advantages and disadvantages. To overcome the disadvantages and improve the strong sides, different oil extraction methods are combined. Moreover, to decrease the environmental impacts of solvents of chemical extraction, different methods of oil extraction have been developed. For instance, Conte, et al. [109] have extracted safflower oil by Soxhlet extraction, ultrasound-assisted extraction (UAE) and pressurized liquid extraction (PLE) techniques (using pressurized ethanol). Soxhlet and ultrasound-assisted extractions gave maximum global oil yield of 36.53% and 30.41%, respectively (70 °C and 240 min) while a maximum global yield for pressurized liquid extraction would be 25.62% [109]. According to the authors, although traditional extraction methods (Soxhlet and UAE) showed maximum global oil yields, the advantages derived from PLE make it a promising alternative for the extraction of essential oil from vegetable matrices due to the reduction of solvent consumption and extraction time.

At optimal conditions of sonication, ultrasonic-assisted extraction (UAE) of raspberry seed oil was able to provide a higher content of beneficial unsaturated fatty acids, whereas conventional Soxhlet extraction resulted in a higher amount of saturated fatty acids [110]. Ultrasound-assisted extraction gave grape seed oil yield (14% w/w) similar to Soxhlet extraction in 6 hours, and no significant differences for the major fatty acids was observed in oils extracted by both methods. The advantage of using ultrasound is that it's lower solvent consumption and a shorter extraction time [111].

6. Future Prospective of Oil Extraction Methods

Biodiesel production from non-edible feedstocks is increasingly attractive alternative to both fossil diesels and renewable fuels derived from food crops. Thus, one of the current research focus in biodiesel production is optimization of oil extraction methods from non-edible oils sources, characterization the oils and suitability test for biodiesel [112], and searching for appropriate methods of biodiesel production from these oils [4,17]. Non-edible biodiesel feedstocks include non-edible oils, animal fats and waste oils [4, 13] and algal biomass [10,11,12]. Some of the recently optimized non-edible seed oil extraction methods include extraction from seeds of waste date pits (*Phoenix dactylifera L.*) [112], Sesame (*Sesamum indicum L.*) [113], jatropha seed kernels [114], beauty leaf seed *(Calophyllum inophyllum)* [85], *Moringa oleifera* [115] and karanja (*Pongamia pinnata*) [116].

According to Sajjadi et al. [17], animal fats are important feedstocks for biodiesel production as their cost is substantially lower than the cost of vegetable oil. However, many types of animal fats contain high amount of saturated fatty acids, which make the transesterification process difficult. To overcome such problems, various biodiesel production methods have been optimized by different investigators. For instance, Kumar and Math [117] investigated the combined effects of catalyst (NaOH) concentration, reaction time and methanol quantity on biodiesel yield from waste animal fat at 55 °C to 60 °C, and obtained the maximum animal fat methyl ester yield of 91% v/v, at 35% v/v methanol and 0.46% w/v catalyst within 90 minutes. Chakraborty and Sahu [118] have also carried out a study on the impacts of methanol to goat tallow molar ratio, infrared radiation assisted reactor (IRAR) temperature and H_2SO_4 concentration on the tallow conversion to biodiesel. Under optimal conditions, 96.7% FFA conversion was achieved within 2.5 h at 59.93 wt.% H_2SO_4, 69.97 °C IRAR temperature and 31.88:1 methanol to tallow molar ratio. According to the authors, infrared radiation strategy could significantly reduce the reaction time compared to conventionally heated reactor while providing appreciably high biodiesel yield. Nuhu and Kovo [119] used a two-step transesterification to produce biodiesel from chicken fat due to the presence of high   FFA (4.16%) in the feedstock, and the first esterification step was a pretreatment process that could reduce the FFA to 0.43%. The second step, the transesterification reaction, yielded 93.4% fat methyl ester from 50g of chicken fat at 60 °C reaction temperature and within 2 hours corresponding to 6:1 molar ratio of oil-to-methanol and 1% wt catalyst concentration.

From various types of biomass, microalgae have the potential of becoming a significant energy source for biofuel production in the coming years. Currently, researches are mainly focusing on optimization of cultivation methods and the conversion of microalgae to biodiesel (lipids for biodiesel production) [120]. Martinez-Guerra and Gude [121] has also wrote that algal biodiesel production will play a significant role in sustaining future transportation fuel supplies, and a large number of researchers around the world are investigating into making this process sustainable by increasing the energy gains and by optimizing resource-utilization efficiencies. Some of the studies that focus on optimization of biodiesel production from microalgae include the investigations by Misau et al. [122], Gülyurt et al. [123], Barreiro, et al. [120] and Rajendran et al. [124].

7. Conclusions

The increasing demand of energy, where the major part of that energy is derived from fossil sources and the problem associated with petroleum fuels have led to search for renewable alternative energy sources of which biodiesel is a promising alternative. The potential feedstock of biodiesel include, edible and non-edible oils, animal fats, waste oils and algal biomass. However, nowadays, more than 95% of the world biodiesel is produced from edible oils and this resulted in food versus fuel debates, rising in the price of oil and environmental problems. To overcome these problems, it is important to use relatively cheaper and non-edible biodiesel feedstock such as non-edible oils, waste animal fats and waste oils.

Many non-edible plat oils have fatty acid composition and other physico-chemical properties that enable them to be suitable for biodiesel production as that of edible oils. Moreover, many potential non-edible plant oil for biodiesel have been identified, and the oil extraction and biodiesel production methods have also been optimized. Methods to extract oil from waste animal fats and refining animal oils and waste oils, and converting them to biodiesel were also optimized by different scholars.

The major oil extraction methods are mechanical extraction, chemical or solvent extraction, and enzymatic extraction. From these methods, chemical or solvent oil extraction method, particularly, Soxhlet extraction by using hexane as solvent, is the most widely used method due to its efficiency of oil extraction. However, chemical oil extraction method has a negative environmental impact. There are also other oil extraction method such as accelerated solvent extraction, supercritical fluid extraction, microwave-assisted extraction and ultrasonic-assisted extraction.

All oil extraction methods have their own advantages and disadvantages. Therefore, by combining the appropriate oil extraction methods, it is possible to reduce the disadvantages and improve the oil extraction efficiency and reduce the negative environmental impacts. Furthermore, for non-edible and low-cost biodiesel feedstocks gradually gain acceptance and well establish and continue to settle in the biodiesel market, various aspects must be scrutinized and studied. Researches that focus on the study of low-cost biodiesel feedstocks, various efficient and environmental-friendly oil extraction techniques, and study of oil yield and fatty acid profiles of non-edible oils, animal fats and waste oils and efficient and cost effective biodiesel conversion technologies are crucial. It can also be concluded that the emphasis must be given to those feedstocks which are neither compete with food crops nor lead to land clearing, and provide significant greenhouse-gas reductions.

Acknowledgments

The authors would like to express their gratitude to the EnPe/NORHED project of Norad at the Norwegian University of Life Sciences, Faculty of Sciences and Technology for their financial support.

Conflict of Interest

All authors declare no conflicts of interest in this paper.

References

1. Khan TMY, Atabani AE, Badruddin IA (2014) Recent scenario and technologies to utilize non-edible oils for biodiesel production. *Renew Sust Energ Rev* 37: 840–851.
2. BP Energy Outlook (2016). BP Energy Outlook 2035, BP PLC. Available from: https://www.bp.com/content/dam/bp/pdf/energy-economics/energy-outlook-2016/bp-energy-out look-2016.
3. Selvakumar MJ, Alexis SJ (2016) Renewable fuel production technologies. *MEJSR* 24: 2502–2509.
4. Bhuiya M, Rasul M, Khan M, et al. (2016) Prospects of 2nd generation biodiesel as a sustainable fuel—Part: 1 selection of feedstocks, oil extraction techniques and conversion technologies. *Renew Sust Energ Rev* 55: 1109–1128.
5. Ho DP, Ngo HH, Guo W (2014) A mini review on renewable sources for biofuel. *Bioresource Technol* 169: 742–749.
6. Atabani AE, Silitonga AS, Ong HC, et al. (2013) Non-edible vegetable oils: a critical evaluation of oil extraction, fatty acid compositions, biodiesel production, characteristics, engine performance and emissions production. *Renew Sust Energ Rev* 18: 211–245.
7. Knothe G (2010) Biodiesel and renewable diesel: a comparison. *Prog Energ Combust* 36: 364–373.
8. Kannahi M, Arulmozhi R (2013) Production of biodiesel from edible and non-edible oils using rhizopus oryzae and aspergillus niger. *Asian J Plant Sci Res* 3: 60–64.
9. Naik SN, Goud VV, Rout PK, et al. (2010) Production of first and second generation biofuels: a comprehensive review. *Renew Sust Energ Rev* 14: 578–597.
10. Lee RA, Lavoie JM (2013) From first-to third-generation biofuels: challenges of producing a commodity from a biomass of increasing complexity. *Animal Front* 3: 6–11.
11. Moser BR (2009) Biodiesel production, properties, and feedstocks. *In Vitro Cell Dev Pl* 45: 229–266.

12. No SY (2011) Inedible vegetable oils and their derivatives for alternative diesel fuels in CI engines: a review. *Renew Sust Energ Rev* 15: 131–149.

13. Atabani AE, Silitonga AS, Badruddin IA, et al. (2012) A comprehensive review on biodiesel as an alternative energy resource and its characteristics. *Renew Sust Energ Rev* 16: 2070–2093.

14. Balat M (2011) Potential alternatives to edible oils for biodiesel production-a review of current work. *Energ Convers Manage* 52: 1479–1492.

15. Marchetti JM (2012) A summary of the available technologies for biodiesel production based on a comparison of different feedstock's properties. *Process Saf Environ* 90: 157–163.

16. Patil PD, Deng S (2009) Optimization of biodiesel production from edible and non-edible vegetable oils. *Fuel* 88: 1302–1306.

17. Sajjadi B, Raman AAA, Arandiyan H (2016) A comprehensive review on properties of edible and non-edible vegetable oil-based biodiesel: composition, specifications and prediction models. *Renew Sust Energ Rev* 63: 62–92.

18. Avhad M, Marchetti J (2015) A review on recent advancement in catalytic materials for biodiesel production. *Renew Sust Energ Rev* 50: 696–718.

19. Ong H, Silitonga A, Masjuki H, et al. (2013) Production and comparative fuel properties of biodiesel from non-edible oils: Jatropha curcas, Sterculia foetida and Ceiba pentandra. *Energ Convers Manage* 73: 245–255.

20. Shahid EM, Jamal Y (2011) Production of biodiesel: a technical review. *Renew Sus Energ Rev* 15: 4732–4745.

21. Çetinkaya M, Ulusoy Y, Tekìn Y, et al. (2005) Engine and winter road test performances of used cooking oil originated biodiesel. *Energ Convers Manage* 46: 1279–1291.

22. Atadashi I, Aroua M, Aziz AA (2010) High quality biodiesel and its diesel engine application: a review. *Renew Sust Energ Rev* 14: 1999–2008.

23. Ahmad A, Yasin NM, Derek C, et al. (2011) Microalgae as a sustainable energy source for biodiesel production: a review. *Renew Sust Energ Rev* 15: 584–593.

24. Lin L, Cunshan Z, Vittayapadung et al. (2011) Opportunities and challenges for biodiesel fuel. *Appl Energ* 88: 1020–1031.

25. Tan K, Lee K, Mohamed A (2011) Potential of waste palm cooking oil for catalyst-free biodiesel production. *Energy* 36: 2085–2088.

26. Silitonga A, Atabani A, Mahlia T, et al. (2011) A review on prospect of *Jatropha curcas* for biodiesel in Indonesia. *Renew Sust Energ Rev* 15: 3733–3756.

27. Juan JC, Kartika DA, Wu TY, et al. (2011) Biodiesel production from jatropha oil by catalytic and non-catalytic approaches: an overview. *Bioresource Technol* 102: 452–460.

28. Lim S, Teong LK (2010) Recent trends, opportunities and challenges of biodiesel in Malaysia: an overview. *Renew Sust Energ Rev* 14: 938–954.

29. Kafuku G, Mbarawa M (2010) Biodiesel production from *Croton megalocarpu*s oil and its process optimization. *Fuel* 89: 2556–2560.

30. Singh S, Singh D (2010) Biodiesel production through the use of different sources and characterization of oils and their esters as the substitute of diesel: a review. *Renew Sust Energ Rev* 14: 200–216.

31. Kibazohi O, Sangwan R (2011) Vegetable oil production potential from *Jatropha curcas, Croton megalocarpus, Aleurites moluccana, Moringa oleifera* and *Pachira glabra*: assessment of renewable energy resources for bio-energy production in Africa. *Biomass Bioenerg* 35: 1352–1356.

32. Supamathanon N, Wittayakun J, Prayoonpokarach S (2011) Properties of *Jatropha seed* oil from Northeastern Thailand and its transesterification catalyzed by potassium supported on NaY zeolite. *J Ind Eng Chem* 17: 182–185.

33. Thiruvengadaravi K, Nandagopal J, Baskaralingam, et al. (2012) Acid-catalyzed esterification of karanja (*Pongamia pinnata*) oil with high free fatty acids for biodiesel production. *Fuel* 98: 1–4.

34. Al Awad AS, Selim MY, Zeibak AF, et al. (2014) Jojoba ethyl ester production and properties of ethanol blends. *Fuel* 124: 73–75.

35. Kumar R, Tiwari P, Garg S (2013) Alkali transesterification of linseed oil for biodiesel production. *Fuel* 104: 553–560.

36. Taufiq YYH, Teo SH, Rashid U, et al. (2014) Transesterification of Jatropha curcas crude oil to biodiesel on calcium lanthanum mixed oxide catalyst: effect of stoichiometric composition. *Energ Convers Manage* 88: 1290–1296.

37. Karmakar A, Karmakar S, Mukherjee S (2010) Properties of various plants and animals feedstocks for biodiesel production. *Bioresource Technol* 101: 7201–7210.

38. Lee S, Radu S, Ariffin A, et al. (2015) Physico-chemical characterization of oils extracted from noni, spinach, lady's finger, bitter gourd and mustard seeds, and copra. *Int J Food Prop* 18: 2508–2527.

39. Demiral İ, Eryazıcı A, Şensöz S (2012) Bio-oil production from pyrolysis of corncob (*Zea mays L.*). *Biomass Bioenerg* 36: 43–49.

40. Da Porto C, Decorti D, Tubaro F (2012) Fatty acid composition and oxidation stability of hemp (*Cannabis sativa L.*) seed oil extracted by supercritical carbon dioxide. *Ind Crop Prod* 36: 401–404.

41. Fadhil AB, Abdulahad WS (2014) Transesterification of mustard (*Brassica nigra*) seed oil with ethanol: purification of the crude ethyl ester with activated carbon produced from de-oiled cake. *Energ Convers Manage* 77: 495–503.

42. Mitra P, Ramaswamy HS, Chang KS (2009) Pumpkin (*Cucurbita maxima*) seed oil extraction using supercritical carbon dioxide and physicochemical properties of the oil. *J Food Eng* 95: 208–213.

43. Alfawaz MA (2004) Chemical composition and oil characteristics of pumpkin (*Cucurbita maxima)* seed kernels. *Food Sci Agr* 2: 5–18.

44. Duz MZ, Saydut A, Ozturk G (2011) Alkali catalyzed transesterification of safflower seed oil assisted by microwave irradiation. *Fuel Process Technol* 92: 308–313.

45. Saydut A, Duz MZ, Kaya C, et al. (2008) Transesterified sesame (*Sesamum indicum L.*) seed oil as a biodiesel fuel. *Bioresource Technol* 99: 6656–6660.

46. Bergmann J, Tupinambá D, Costa O, et al. (2013) Biodiesel production in Brazil and alternative biomass feedstocks. *Renew Sust Energ Rev* 21: 411–420.

47. Phani RC, Chaitanya DR (2014) A study on Sunflower oil quality in different seasons. *Afro Asian J Sci Tech* 1: 176–177.

48. Quampah A, Huang ZR, Wu JG, et al. (2012) Estimation of oil content and fatty acid composition in cottonseed kernel powder using near infrared reflectance spectroscopy. *J Am Oil Chem Soc* 89: 567–575.

49. Gui MM, Lee K, Bhatia S (2008) Feasibility of edible oil vs. non-edible oil vs. waste edible oil as biodiesel feedstock. *Energy* 33: 1646–1653.

50. Azam MM, Waris A, Nahar N (2005) Prospects and potential of fatty acid methyl esters of some non-traditional seed oils for use as biodiesel in India. *Biomass Bioenerg* 29: 293–302.

51. Balat M, Balat H (2010) Progress in biodiesel processing. *Appl Energ* 87: 1815–1835.

52. Oliveira JFG, Lucena IL, Saboya RMA, et al. (2010) Biodiesel production from waste coconut oil by esterification with ethanol: the effect of water removal by adsorption. *Renew Energ* 35: 2581–2584.

53. Li SY, Stuart JD, Li Y, et al. (2010) The feasibility of converting *Cannabis sativa L.* oil into biodiesel. *Bioresource Technol* 101: 8457–8460.

54. Ramos MJ, Fernández CM, Casas A, et al. (2009) Influence of fatty acid composition of raw materials on biodiesel properties. *Bioresource Technol* 100: 261–268.

55. Zhang Y, Wong WT, Yung KF (2013) One-step production of biodiesel from rice bran oil catalyzed by chlorosulfonic acid modified zirconia via simultaneous esterification and transesterification. *Bioresource Technol* 147: 59–64.

56. Mihaela P, Josef R, Monica N, et al. (2013) Perspectives of safflower oil as biodiesel source for South Eastern Europe (comparative study: Safflower, soybean and rapeseed). *Fuel* 111: 114–119.

57. Park YW, Chang PS, Lee J (2010) Application of triacylglycerol and fatty acid analyses to discriminate blended sesame oil with soybean oil. *Food Chem* 123: 377–383.

58. Ghanei R, Moradi G, Taherpour KR, et al. (2011) Variation of physical properties during transesterification of sunflower oil to biodiesel as an approach to predict reaction progress. *Fuel Process Technol* 92: 1593–1598.

59. Bhuiya M, Rasul M, Khan M, et al. (2016) Prospects of 2nd generation biodiesel as a sustainable fuel-Part 2: properties, performance and emission characteristics. *Renew Sust Energ Rev* 55: 1129–1146.

60. Athalye S, Sharma SR, Peretti S, et al. (2013) Producing biodiesel from cottonseed oil using Rhizopus oryzae whole cell biocatalysts: culture media and cultivation period optimization. *Energ Sust Dev* 17: 331–336.

61. Pinzi S, Garcia I, Lopez GF, et al. (2009) The ideal vegetable oil-based biodiesel composition: a review of social, economical and technical implications. *Energ Fuel* 23: 2325–2341.

62. Fattah IR, Masjuki H, Liaquat A, et al. (2013) Impact of various biodiesel fuels obtained from edible and non-edible oils on engine exhaust gas and noise emissions. *Renew Sust Energ Rev* 18: 552–567.

63. Knothe G, Steidley KR (2005) Kinematic viscosity of biodiesel fuel components and related compounds. Influence of compound structure and comparison to petrodiesel fuel components. *Fuel* 84: 1059–1065.

64. Hincapié G, Mondragón F, López D (2011) Conventional and in situ transesterification of castor seed oil for biodiesel production. *Fuel* 90: 1618–1623.

65. Joshi H, Moser BR, Shah SN, et al. (2010) Improvement of fuel properties of cottonseed oil methyl esters with commercial additives. *Eur J Lipid Sci Tech* 112: 802–809.

66. Okia C, Kwetegyeka J, Okiror P, et al. (2013) Physico-chemical characteristics and fatty acid profile of desert date kernel oil in Uganda. *Afr Crop Sci J* 21: 723–734.

67. Shah SN, Sharma BK, Moser BR, et al. (2010) Preparation and evaluation of jojoba oil methyl esters as biodiesel and as a blend component in ultra-low sulfur diesel fuel. *Bioenerg Res* 3: 214–223.

68. Sahoo P, Das L (2009) Process optimization for biodiesel production from Jatropha, Karanja and Polanga oils. *Fuel* 88: 1588–1594.

69. Demirbas A (2009) Production of biodiesel fuels from linseed oil using methanol and ethanol in non-catalytic SCF conditions. *Biomass Bioenerg* 33: 113–118.

70. Jena PC, Raheman H, Kumar GP, et al. (2010) Biodiesel production from mixture of mahua and simarouba oils with high free fatty acids. *Biomass Bioenerg* 34: 1108–1116.

71. Kafuku G, Mbarawa M (2010) Alkaline catalyzed biodiesel production from *Moringa oleifera* oil with optimized production parameters. *Appl Energ* 87: 2561–2565.

72. Betiku E, Omilakin OR, Ajala SO, et al. (2014) Mathematical modeling and process parameters optimization studies by artificial neural network and response surface methodology: a case of non-edible neem (*Azadirachta indica*) seed oil biodiesel synthesis. *Energy* 72: 266–273.

73. Satyanarayana M, Muraleedharan C (2011) A comparative study of vegetable oil methyl esters (biodiesels). *Energy* 36: 2129–2137.

74. Parlak A, Ayhan V, Cesur İ, et al. (2013) Investigation of the effects of steam injection on performance and emissions of a diesel engine fuelled with tobacco seed oil methyl ester. *Fuel Process Technol* 116: 101–109.

75. Al Hamamre Z, Yamin J (2014) Parametric study of the alkali catalyzed transesterification of waste frying oil for biodiesel production. *Energ Convers Manage* 79: 246–254.

76. Öner C, Altun S (2009) Biodiesel production from inedible animal tallow and an experimental investigation of its use as alternative fuel in a direct injection diesel engine. *Appl Energ* 86: 2114–2120.

77. Gürü M, Koca A, Can Ö, et al. (2010) Biodiesel production from waste chicken fat based sources and evaluation with Mg based additive in a diesel engine. *Renew Energ* 35: 637–643.

78. Lu J, Nie K, Xie F, et al. (2007) Enzymatic synthesis of fatty acid methyl esters from lard with immobilized Candida sp. *Process Biochem* 42: 1367–1370.

79. Diaz FW, Riley MR, Zimmt W, et al. (2009) Pretreatment of yellow grease for efficient production of fatty acid methyl esters. *Biomass Bioenerg* 33: 558–563.

80. Adewale P, Dumont MJ, Ngadi M (2015) Recent trends of biodiesel production from animal fat wastes and associated production techniques. *Renew Sust Energ Rev* 45: 574–588.

81. Banković IIB, Stamenković OS, Veljković VB (2012) Biodiesel production from non-edible plant oils. *Renew Sust Energ Rev* 16: 3621–3647.

82. Jahirul MI, Brown JR, Senadeera W, et al. (2013) Optimisation of bio-oil extraction process from beauty leaf (*Calophyllum inophyllum*) oil seed as a second generation biodiesel source. *Procedia Eng* 56: 619–624.

83. Achten W, Verchot L, Franken YJ, et al. (2008) Jatropha bio-diesel production and use. *Biomass Bioenerg* 32: 1063–1084.

84. Ramachandran S, Singh SK, Larroche C, et al. (2007) Oil cakes and their biotechnological applications-a review. *Bioresource Technol* 98: 2000–2009.

85. Bhuiya M, Rasul M, Khan MMK, et al. (2015) Optimisation of oil extraction process from Australian native beauty leaf seed (*Calophyllum Inophyllum*). *Energ Procedia* 75: 56–61.

86. Forson F, Oduro E, Hammond DE (2004) Performance of jatropha oil blends in a diesel engine. *Renew Energ* 29: 1135–1145.

87. Lokanatham RPRK (2013) Extraction and use of non-edible oils in bio-diesel preparation with performance and emission analysis on C.I. engine. *Int J Eng Res Dev* 6: 35–45.

88. Beerens P (2007) Screw-pressing of Jatropha seeds for fuelling purposes in less developed countries. Eindhoven University of Technology Ministerio de Ambiente Energía-MINAE-(2007)"Plan Nacional de Biocombustibles", Costa Rica. Available from: http://citeseerx.ist.psu.edu/viewdoc/download?doi=10.1.1.454.2241&rep=rep1&type=pdf.

89. Sayyar S, Abidin ZZ, Yunus R, et al. (2009) Extraction of oil from Jatropha seeds-optimization and kinetics. *Am J Appl Sci* 6:1390–1395.

90. Banat F, Pal P, Jwaied N, et al. (2013) Extraction of olive oil from olive cake using soxhlet apparatus. *Am J Oil Chem Technol* 1: 2326–6570.

91. Mahanta P, Shrivastava A (2004) Technology development of bio-diesel as an energy alternative. Department of Mechanical Engineering, Indian Institute of Technology. Available from: http://www.newagepublishers.com/samplechapter/001305.pdf.

92. Khattab RY, Zeitoun MA (2013) Quality evaluation of flaxseed oil obtained by different extraction techniques. *LWT Food Sci Techno* 53: 338–345.

93. Sarip MSM, Morad NA, Yamashita Y, et al. (2016) Crude palm oil (CPO) extraction using hot compressed water (HCW). *Sep Purif Technol* 169: 103–112.

94. Dunford NT, Zhang M (2003) Pressurized solvent extraction of wheat germ oil. *Food Res Int* 36: 905–909.

95. Oomah BD, Sitter L (2009) Characteristics of flaxseed hull oil. *Food Chem* 114: 623–628.

96. Rosenthal A, Pyle D, Niranjan K, et al. (2001) Combined effect of operational variables and enzyme activity on aqueous enzymatic extraction of oil and protein from soybean. *Enzyme Microb Tech* 28: 499–509.

97. Shah S, Sharma A, Gupta M (2005) Extraction of oil from *Jatropha curcas L*. seed kernels by combination of ultrasonication and aqueous enzymatic oil extraction. *Bioresource Technol* 96: 121–123.

98. Augustus G, Jayabalan M, Seiler G (2002) Evaluation and bioinduction of energy components of *Jatropha curcas*. *Biomass Bioenerg* 23: 161–164.

99. Shah S, Sharma A, Gupta M (2004) Extraction of oil from *Jatropha curcas L*. seed kernels by enzyme assisted three phase partitioning. *Ind Crop Prod* 20: 275–279.

100. Akanda MJH, Sarker MZI, Ferdosh S, et al. (2012) Applications of supercritical fluid extraction (SFE) of palm oil and oil from natural sources. *Molecules* 17: 1764–1794.

101. Maran JP, Priya B (2015) Supercritical fluid extraction of oil from muskmelon (*Cucumis melo*) seeds. *J Taiwan Inst Chem E* 47: 71–78.

102. Rubio RN, Sara M, Beltrán S, et al. (2008) Supercritical fluid extraction of the omega-3 rich oil contained in hake (*Merluccius capensis-Merluccius paradoxus*) by-products: study of the influence of process parameters on the extraction yield and oil quality. *J Supercrit Fluid* 47: 215–226.

103. Rubio RN, Sara M, Beltrán S, et al. (2012) Supercritical fluid extraction of fish oil from fish by-products: a comparison with other extraction methods. *J Food Eng* 109: 238–248.

104. Hao JY, Han W, Xue BY, et al. (2002) Microwave-assisted extraction of artemisinin from *Artemisia annua L*. *Sep Purif Technol* 28: 191–196.

105. Cardoso UGA, Juárez BGP, SosaMorales ME, et al. (2013) Microwave-assisted extraction of essential oils from herbs. *J Microwave Power E E* 47: 63–72.

106. Jiao J, Li ZG, Gai QY, et al. (2014) Microwave-assisted aqueous enzymatic extraction of oil from pumpkin seeds and evaluation of its physicochemical properties, fatty acid compositions and antioxidant activities. *Food Chem* 147: 17–24.

107. Kanitkar A, Sabliov C, Balasubramanian S, et al. (2011) Microwave-assisted extraction of soybean and rice bran oil: yield and extraction kinetics. *T Asabe* 54: 1387–1394.

108. Ali M, Watson IA (2014) Comparison of oil extraction methods, energy analysis and biodiesel production from flax seeds. *Int J Energ Res* 38: 614–625.

109. Conte R, Gullich LM, Bilibio D, et al. (2016) Pressurized liquid extraction and chemical characterization of safflower oil: a comparison between methods. *Food Chem* 213: 425–430.

110. Teng H, Chen L, Huang Q, et al. (2016) Ultrasonic-assisted extraction of raspberry seed oil and evaluation of its physicochemical properties, fatty acid compositions and antioxidant activities. *Plos One* 11: In press.

111. Boey PL, Ganesan S, Maniam GP, et al. (2011) Ultrasound aided in situ trasnesterification of crude palm oil adsorbed on spent bleachin clay. *Energ Convers Manage* 52: 2081–2084.

112. Jamil F, Ala'a H, Al Haj L, et al. (2016) Optimization of oil extraction from waste "Date pits" for biodiesel production. *Energ Convers Manage* 117: 264–272.

113. Elkhaleefa A, Shigidi I (2015) Optimization of sesame oil extraction process conditions. *Adv Chem Eng Sci* 5: 305.

114. Yahaya S, Giwa SO, Ibrahim M, et al. (2016) Extraction of oil from Jatropha seed kernels: optimization and characterization. *Int J Chem Tech Res* 9: 758–770.

115. Bhutada PR, Jadhav AJ, Pinjari DV, et al. (2016) Solvent assisted extraction of oil from *Moringa oleifera Lam.* seeds. *Ind Crop Prod* 82: 74–80.

116. Sagwan S, Rao D, Sharma R (2016) Comparative physical properties of karanj seed oil by using different organic solvents: an environmental viable fuel. *Int J Adv Biotechnol Res* 1: 540–544.

117. Kumar NB, Math MC (2016) Application of response surface methodology for optimization of biodiesel production by transesterification of animal fat with methanol. *Int J Renew Energ Res* 6: 74–79.

118. Chakraborty R, Sahu H (2014) Intensification of biodiesel production from waste goat tallow using infrared radiation: process evaluation through response surface methodology and artificial neural network. *Appl Energ* 114: 827–836.

119. Nuhu S, Kovo A (2015) Production and characterization of biodiesel from chicken fat. *Scholarly J Agr Sci* 5: 22–29.

120. Barreiro DL, Prins W, Ronsse F, et al. (2013) Hydrothermal liquefaction (HTL) of microalgae for biofuel production: state of the art review and future prospects. *Biomass Bioenerg* 53: 113–127.

121. Martinez GE, Gude VG (2016) Energy aspects of microalgal biodiesel production. *AIMS Energ* 4: 347–362.

122. Misau IM, Mohammed UA, Ahmed SI (2016) Optimization of reaction condition for biodiesel production from microalgae. *J Sci Res Stud* 3: 1–5.

123. Gülyurt MÖ, Özçimen D, İnan B (2016) Biodiesel production from *Chlorella protothecoides* oil by microwave-assisted transesterification. *Int J Mol Sci* 17: 579.

124. Rajendran R, Kanimozhi B, Prabhavathi P, et al. (2015) A method of central composite design (CCD) for optimization of biodiesel production from *Chlorella vulgaris*. *J Petrol Environ Eng* 6: In press.

PERMISSIONS

LIST OF CONTRIBUTORS

Gul Ahmad Ludin and Tomonobu Senjyu
Electrical and Electronics Engineering Department, University of the Ryukyus, Okinawa, Japan

Mohammad Amin Amin and Assadullah Aminzay
Electrical Power Engineering Department, Kabul Polytechnic University, Kabul, Afghanistan

Ankita Juneja
Biological and Ecological Engineering, Oregon State University, Corvallis OR 97331, USA Agricultural and Biological Engineering, University of Illinois Urbana Champaign, Urbana IL 61801, USA

Ganti S. Murthy
Biological and Ecological Engineering, Oregon State University, Corvallis OR 97331, USA

Mulualem T. Yeshalem and Baseem Khan
School of Electrical & Computer Engineering, Hawassa University Institute of Technology, Hawassa, 05, Ethiopia

Albert K. Awopone and Ahmed F. Zobaa
Department of Electronic and Computer Engineering, Brunel University London, Uxbridge UB8 3PH, UK

Shahrouz Abolhosseini
National Iranian Oil Company, International Affairs, Petroleum University of Technology, Teheran, Iran

Almas Heshmati
Sogang University, Seoul, Korea

Masoomeh Rashidghalam
University of Tabriz, Tabriz, Iran

Kamil Dino Adem and Demiss Alemu Ambie
School of Mechanical and Industrial Engineering, Addis Ababa Institute of Technology, Addis Ababa University, Addis Ababa, Ethiopia

Takvor Soukissian and Hara Kyriakidou
Institute of Oceanography, Hellenic Centre for Marine Research, Anavyssos, Greece

Anastasios Papadopoulos
Institute of Marine Biological Resources and Inland Waters, Hellenic Centre for Marine Research, Anavyssos, Greece

Panagiotis Skrimizeas, Evripides Avgoustoglou, Antigoni Voudouri and Flora Gofa
Hellenic National Meteorological Service, Hellinikon, Athens, Greece

Flora Karathanasi
Institute of Oceanography, Hellenic Centre for Marine Research, Anavyssos, Greece Department of Naval Architecture and Marine Engineering, National Technical University of Athens, Zografos, Athens, Greece

Panagiotis Axaopoulos
Department of Geology, Centre for Arctic Gas Hydrate, Environment and Climate, UiT, The Arctic University of Norway, Norway

Christos Tsalis
Atmospheric Modeling and Weather Forecasting Group, Division of Applied Physics, School of Physics, University of Athens, Athens, Greece

Petros Katsafados
Department of Geography, Harokopio University, Athens, Greece

Pedro J. B. F. N. Beirão
Instituto Politécnico de Coimbra, ISEC, DEM, Coimbra, Portugal
LAETA, IDMEC, Instituto Superior Técnico, Universidade de Lisboa, Lisboa, Portugal

Cândida M. S. P. Malça and Raimundo P. Felismina
Instituto Politécnico de Coimbra, ISEC, DEM, Coimbra, Portugal

María del P. Pablo-Romero
Department of Economic Analysis and Political Economy, Faculty of Economics and Business
Sciences, University of Seville, Ramon y Cajal 1, 41018 Seville, Spain

Rafael Pozo-Barajas
Department of Financial Economy and Operations Management, Faculty of Economics and Business Sciences, University of Seville, Ramon y Cajal 1, 41018 Seville, Spain

Angel G. Gonzalez-Rodriguez
Department of Electronic Engineering and Automation, University of Jaen, Jaen, Spain

Manuel Burgos Payan, Jesús Riquelme Santos and Javier Serrano Gonzalez
Department of Electrical Engineering, University of Seville, Camino de los Descubrimientos s/n, Seville, Spain

Vo Thanh Phuoc and Kunio Yoshikawa
Yoshikawa Laboratory, Department of Environmental Science and Technology, Tokyo Institute of Technology, Yokohama, Kanagawa, 226-8502, Japan

Yadessa Gonfa Keneni and Jorge Mario Marchetti
Faculty Sciences and Technology, Norwegian University of Life Sciences, Drøbakveien 31, Ås 1432, Norway

Index

A

Accelerated Solvent Extraction (ASE), 206, 208

Aegean And Ionian Seas, 101-102, 104-108, 110-117, 119

Afghan Energy System, 4-5

Algae Cultivation, 24

Algae Strain, 22

Alternative Energy Sources, 74-75, 78, 196-197, 212

Ase Accelerated Solvent Extraction, 196

Astm American Society of Testing and Materials, 87

B

Base Transceiver Station (BTSS), 39

Bio-crude Production, 26

Biodiesel Feedstock, 196, 203, 212, 216

Biomass, 1-3, 5, 8, 12-14, 17, 19, 21-22, 36-38, 40, 57, 87-90, 94, 97-100, 152-154, 193-195, 197-198, 205, 211-213, 215-219

Bound Fatty Acids (BFAS), 181-182

Buoy Measurements, 101, 110

C

Carbon Taxes, 56, 65-66

Chp (Combined Heat and Power Generation), 24

Clean Development Mechanism (CDM), 60, 87, 97

Climate and Energy Policy, 138

Continuous Evolutionary Algorithm, 161

Cost-benefit Analysis, 67, 103

D

Dabs (Da Afghanistan Breshna Sherkat), 5

Diesel Generator (DG) System, 39, 41

Diversification Policy, 72

Dynamic Model and Simulator, 127

E

Economic Evaluation, 39, 67, 103

Edible Oil, 196-197, 200, 203, 216

Emission Trading (ET), 60

Energy Security, 4, 17, 65, 67, 71-76, 78-86, 143, 149, 154, 159

Energy Security Strategy, 78

Energy Supply and Demand, 74-75

Environmental-friendly, 196, 208-209, 213

Enzymatic Extraction Methods, 196, 208-209

Esterification, 181-183, 185-187, 189-192, 194, 212, 215-216

F

Fatty Acid, 181-184, 186, 188, 191, 193-194, 196-198, 201-205, 208, 212-213, 215-217, 219

Fatty Acids (TFAS), 181, 186, 188-189, 192

Finite Element Analyses (FEA), 123

Free Fatty Acids (FFAS), 181-182, 186, 188

G

Gasifier Stove, 87, 89-90, 94, 99

Geopolitical Energy Supplier, 80

Geothermal, 1, 3, 5, 8, 15-17, 19, 57, 141

Gradient-based Local Search, 161

Greenhouse Gases (GHG), 56, 58

H

Hybrid Optimization Model For Electric Renewable (HOMER), 39

Hydro Energy, 13

Hydrochar, 181-192

Hydrothermal Carbonization, 181-182, 184, 188, 193-195

Hydrothermal Gasification, 23-24, 27, 29, 37

Hydrothermal Liquefaction, 20-24, 27-28, 35-37, 182, 194-195, 219

Hydrotreating of Htl Biocrude, 27

I

Indoor Air Pollution, 87-89, 92, 94-96, 99-100

Internal Rate of Return (IRR), 161, 167

International Stability, 72, 74

Irar Infrared Radiation Assisted Reactor, 197

J

Joint Comprehensive Plan of Action (JCPOA), 74

Joint Implementation (JI), 60

L

Life Cycle Assessment, 20-22, 35, 37

Long-range Energy Alternatives Planning (LEAP), 56, 58, 70

Long-term Hindcast Simulations, 102

M

Mechanical Oil Extraction, 206

Micro-gasifier, 87-95, 97-99

Microalgae, 21, 24, 31, 35-38, 181-194, 200, 205, 212, 214, 219

Microwave-assisted Extraction (MAE), 206, 209

N

Net Energy Ratio (NER), 28

Net Energy Value (NEV), 28

Non-discrete Evolutionary Algorithm, 161, 163, 167, 176

Non-edible Oil, 196-197, 203, 216

O

Offshore Wind Farms (OWFS), 102, 161

Offshore Wind Potential, 102, 120

Oil Extraction, 27, 36, 182, 193, 196, 198, 205-213, 215, 217-219

Open Source Energy Modelling System (OSEMO-SYS), 56, 58

Optimal Configuration, 42, 49, 161

Optimization, 18, 39, 41, 49-50, 54-55, 58, 71, 121, 136, 161-163, 165, 167, 170-171, 173, 179-180 194, 211-212, 214, 216-220

Optimum Generation Scenarios, 56

P

Performance Evaluation, 87, 91 100

Photovoltaic (PV), 39, 70

R

Reference (REF) Scenario, 60

Regional Security, 73-74, 81-82, 86

Regular Patterns, 161

Renewable Diesel, 20-22, 24 27-29, 31-32, 34-37, 213

Renewable Energy, 1-5, 8-9, 16-21 37, 40-41, 47, 52, 55-59, 61, 63-64 66-70, 72, 75, 85, 105, 123-124 139, 142-144, 147-148, 150 152-155, 157-160, 180, 196, 215

S

Shoreline Transition, 161, 172

Social Benefits, 103

Solar Energy, 1, 8-9, 17, 38, 42-43 51, 54, 57, 70

Stand-alone Systems, 39

Stockholm Environmental Institute (SEI), 58

Supercritical Fluid Extraction (SFE) 206, 208-209, 218

Sustainable Power Generation 56-57

T

Technical Assumptions, 23

Techno-economic Analysis, 20-23 31, 35-36, 38, 41, 55

Theoretical Potential, 1, 8, 11, 13 17

Transesterification, 181-183 185-187, 189-194, 198, 201, 205 211-212, 215-217, 219-220

Transmission And Distribution, 2, 5, 17, 56, 59-61, 66-71

Tridimensional Computer Aided Design (3D CAD), 123, 125

U

Un Framework on Climate Change (UNFCCC), 138

Unfccc United Nation Framework For Climate Change Convention 88

Uv-vis Spectrometer Model, 184

W

Wastewater Treatment, 20-22, 36

Wave Energy Converter, 123, 125 136-137

Wind Energy, 9-11, 19, 41, 43, 45 55, 63, 101-105, 107, 119-122 138-145, 147-149, 151-158 161-162, 180

Wind Power Density, 102, 105-108 111, 114-117, 119

Wind Power System, 39

Wind Resource Assessment Studies, 103

Wind Speed-wind Direction, 102

Worldwide Regional Analysis, 138

Lightning Source UK Ltd.
Milton Keynes UK
UKHW05n1411280518
323150UK00003BB/187/P